Any profit from the sales of *This Tractor Life* will be used to complete the construction of the wood-fired oven at Oranje Tractor Farm. This means that you may get to enjoy one of Murray's fabulous pizzas sooner, rather than later! If you visit once the oven is fired up and Murray is pumping out the pizzas make sure you let us know that you've purchased a book and we'll include a special bonus.

Me & Chris

Chris

Chris & Sharon

Dedication

To my little brother Christopher
whose enthusiasm for life I shall always miss,
and his wife Sharon whose courage and strength I will always admire.

Published in 2014 by Pamela J Lincoln
198 Link Road
Marbelup WA 6330
Australia

Text copyright © Pamela J Lincoln 2014
Photography © Krysta Guille 2013, Pamela J Lincoln 2014

All rights reserved. No part of this publication may be reproduced, stored in a retrieval system or transmitted in any form or by any means, electronic, mechanical, photocopying, recording or otherwise, without the prior permission in writing of the copyright owner.

The moral rights of the author have been asserted.

National Library of Australia
Cataloguing-in-Publication data

Author: Lincoln, Pamela J (Pamela Jane). 1961 - author

Title: This Tractor Life: A memoir of food, wine and woofers at Oranje Tractor Farm/Pamela J Lincoln

ISBN: 978-0-9941528-0-0

Notes: Includes index.

Subjects: Oranje Tractor Farm (Albany, W.A.) – Cooking.
Vineyards – Western Australia – Albany.
Wine and wine making wineries and wine – Western Australia – Albany.
Organic farming – Western Australia – Albany.
Casual labor – Western Australia – Albany.

Dewey Number: 641.5

pam@oranjetractor.com

www.oranjetractor.com

Design: Rebecca Weadon, Croker Lacey Graphic Design
Editing: Penny Simpson, Collyn Gawned, Louise Everett
Photography: Krysta Guille, Pamela Lincoln

Proudly printed by Scott Print (Certified to ISO 14001) in Western Australia
using elemental chlorine free paper sourced from well managed forests.

Acknowledgements

I am forever indebted to Murray and his patience with me, my ideas and my foibles. He not only entertains and embraces but also actively adopts and even promotes the majority of my projects – including this one. Some have borne fruit; others remain seeds with just the potential for germination. We have been able to pour our creative energy into most and nurture the resulting prodigies. Murray is more than the sum of husband and business partner.

All throughout our Oranje Tractor journey, we have been supported by a vast array of friends and family. From planting trees and vines, to hammering nails into weatherboards, harvesting grapes and olives, stomping grapes and staying up late to bottle wine, these unofficial woofers* have been crucial to our survival and success.

Shirley – Murray's mum – stands in a league of her own. Not only is she the only one left standing out of the parental pack, she is also unstoppable. Despite her increasing age Shirley provides an almost immeasurable amount of help. Everything from the cake-cook extraordinaire, to the garden queen, dishwasher, pricing advisor and sous-chef, Shirley pitches in each and every weekend. As the connoisseur of cake and coffee around town, she is a great barometer of what people are eating and readily shares her local experiences.

Murray's dad Barry, or Bazza as we liked to call him amongst ourselves, was – until his untimely death – our primary vineyard hand. In the earlier years, he juggled his beef cattle farm responsibilities with most of our vineyard duties. Later, when he leased the land to a neighbouring farmer, Barry became even more involved in our establishment. The shock of his passing was only surpassed by the shock of how much extra work we needed to do to keep on top of things around the Oranje Tractor Farm without him around.

My parents, too, were stalwarts in the establishment of our home and our vineyard and only just managed to witness the fruits of our collective labour before their premature deaths. My dad once confessed to me that he originally thought we were 'crackers' trying to turn a whole pile of recycled junk into a home, but that he was happy to have been proved wrong. They would regularly make the trek down from Perth to assist in tasks as varied as cleaning bricks, removing paint from timber, planting vine cuttings, and picking and stomping our first few vintages of grapes.

Singling out friends who have pitched in is fraught, I know, with the danger of leaving someone off the list – so apologies in advance, dear friends if you're not on the list I have compiled of regular visitors to Oranje Tractor Farm over the years: Adrian, Anthony, Barb, Billie, Christine, Debra, Don, Erica, Ev, Glenn, Jeck, Jenelle, Jon, the Kims, the LARDS (my former university mates), Laura, Max, Nix, Rio, Renee, Terry and Whitey. It means a lot to me that you've taken time away from your busy lives to stay with us, help us and even entertain us while we pursue our dream. And, to the many friends who have given moral support in this journey, I am sincerely grateful for your presence in this tractor life. You are too many to name, but you know who you are.

We have also been well supported by our local community, particularly with the annual grape harvest. A large number of community groups have contributed their time in return for a donation for either their own cause or a charitable one. Sporting clubs, schools, Army Reserve units, book clubs, as well as Albany's infamous Granny Grommets and more have all made a mark on us.

There are many people who inspire me with either their writing or their cooking, including the journalist Michael Pollan, authors Harold McGee and Michael Symons, and Australia's best loved cook Maggie Beer, and I would like to acknowledge their passion and talent.

One of our many ex-woofers, Priscilla, warrants special mention here as she was the one who introduced me to the fabulous *Hautes Etude de Gout* Gastronomy Diploma, which I completed in 2011. The book *This Tractor Life* came about as a consequence of that undertaking, because it was necessary to prepare a thesis. Mine was titled 'Bonzer Tucker, Mate: An exploration of the opportunity provided by WWOOF as a means of transferring gastronomic culture between hosts and guests, in the Australian context' and was thus the genesis of this book.

I feel it is important to also acknowledge the Noongar people – our local aboriginal inhabitants upon whose land we reside. They are the traditional custodians of this land, and their customs and rituals, including those related to food, are rich and varied.

*woofers is the nickname for people staying with us under the Willing Workers On Organic Farms (WWOOF) host scheme.

Pozible Supporters

Pozible, the crowd-sourcing platform, has been the enabling factor in producing *This Tractor Life*. It has allowed friends, acquaintances and strangers alike to contribute support by way of a donation to the project in what can only be described as an act of faith. Without the faith shown by the following people, this book simply would not exist. Thank you all for your pledges of support, no matter how big or small, and for your patience with the longer-than-expected production time. I hope you hold your head high, are proud of being early adopters of this great concept called crowd-funding, and enjoy a warm fuzzy feeling knowing you were vital to *This Tractor Life*. And I hope you enjoy reading the stories and trying the recipes as much as I have enjoyed sharing them.

Golden Tractor Seat Supporters

Justin Yeung

Silver Tractor Seat Supporters

Glenn Cardwell
Marianne Svahn

Oranje Tractor Seat Supporters

Alison Steer, Anna Pilkington, Anne Jones, Annie Lincoln, Ashley Carruthers, Barbara Madden, Caroline Langston, Cheryl Avery, Christian Wise, Claire Bateman, Danelle Ward, David Pilkington, Don Huffer, Donna Cameron, Glenn Huffer, Ingrid Storm, Jacqui Grieve, Jane Scott, Jarrod Pegler, Juliet Bateman, Kate Thomas, Kim Clark, Kim Buttfield, Lesley Hart, Libby Foster, Louise Everett, Marie O'Dea, Marisina Papalia, Meredith Healy, Michelle Gooding, Nadine Wright Toussaint, Nan and David Anderson, Natalina Cherubino, Rebecca Stephens, Richard Keeler, Rob Hornsby, Robyn Miller, Sarah Bowles, Sarah Liddiard, Sheryl Stephens, Shirley Gomm, Steve Jones, Terreen Stenvers, Warrick Welsh.

Thank you

Woofers

To woofers, one and all; you have enriched our lives by contributing not only your help but also your friendship. Our existence would be infinitely less interesting without having spent time with you. We have enjoyed sharing our home, our dreams and our experience with such amazing people. This book encapsulates mere snippets of our shared time, for it is impossible to describe the depth and breadth of what it has meant to us. Thank you.

Editors

Penny Simpson provided superb editorial skills and I am extremely grateful for her input. In particular, Penny's attention to detail and very short turn-around times are astonishingly good. Murray Gomm and Collyn Gawned also provided feedback on earlier drafts and I thank them sincerely. My thanks also go to Louise Everett and Yann Toussaint, who cast their experienced eyes over the recipe sections and provided significant help. However, any errors, omissions, mistakes or blunders of any kind are purely my responsibility.

Photographers

Krysta Guille's excellent photography skills will be evident to anyone picking up this book. I have always been impressed with her work and am sincerely thankful for her ability to capture the moment. Ulla and Nora, two professional photographers who woofed with us and my niece Rebecca Lincoln, also took many beautiful photographs but sadly, there is not enough space to display much of their work. Thank you anyway. And finally, thanks to Rob Frith, from Acorn Photography, who took the iconic 'smiling tractor' photo that appears here and plenty of other places too. I take full responsibility for the food photographs and trust you, the reader, will understand that I am a rank amateur in this regard.

Graphic Design

Rebecca Weadon is not only a very talented woman; she is a beautiful, kind-hearted soul who is a privilege to know. Her patience and generosity with me and this project is appreciated more than I can express.

Foreword

In March 2006, the organisers of one of the first Great Southern Taste festivals asked me to cook a long table lunch. It was to be held in a marquee in the vineyard at Oranje Tractor and I was the only chef. I was concerned about the protection from the elements that the marquee would provide, so felt a little tentative. "What if the day is a scorcher?" I asked. "Don't worry", was the response, "it never gets that hot down here".

The day arrived, and so did the 80 or so guests. It was 35 degrees in the shade and getting hotter, with no breeze to flush the hot air from the marquee. It could have been a recipe for disaster, but it wasn't. Many of Pam and Murray's mates had travelled from Perth or dropped in from Albany to set up, pour the wines, and to help me cook, plate and serve the menu...all with a smile and careful attention to detail. From the moment the first glasses of Pam's exciting 2004 Riesling were poured it became a celebration of the food of the region matched with Oranje Tractor's wonderful wines, albeit with an extra bead of sweat or two trickling down the faces of the merry crowd.

That day, to me, and my visits since have demonstrated the clear difference between the Oranje Tractor experience and that of so many other wineries. It is visiting a 'community' where tasting quality wine made with passion and commitment is only part of the experience. It is the journey from the gravel car park towards the amazing organic orchard of healthy fruit trees weighed down with whatever is in season. Then down through the herb garden, past one or two bicycles from Murray's collection to the cellar door with its beautiful stained glass rendition of the famous tractor, made by Pam's oldest friend Helen. Then to the rustic open dining space past circular recycled brick garden beds brimming with herbs and annuals, some of those flowers picked to decorate the tables, where platters of fabulous rustic food are served. There you'll probably find one of their Perth mates pitching in to help for the weekend, or a French, German, Korean woofer serving wonderful goodies from the kitchen to locals and tourists passing through. As always a sense of fun, family, and organic purpose pervades every corner of the property.

So to be asked to write the foreword to Tractor Life is a pleasure. Pam's prose reveals her passion for the establishment and unlocks many of the stories that provide the reader with an understanding of it. The recipes offer satisfying food from the hearts and minds of Oranje Tractor folk and are simple, authentic and... bloody tasty!

Russell Blaikie

Chef-Proprietor
Must Winebar Perth
Muster bar & grill Margaret River

Contents

Acknowlegements........... 7

Foreword 11

Introductions 15

History........................ 16

Me.............................. 19

Him 23

Them 25

His Majesty.................. 33

Home.......................... 35

Recipe Notes................ 39

Spring......................... 41

Spring Recipes 49

Summer....................... 79

Summer Recipes........... 85

Autumn 113

Autumn Recipes 119

Winter....................... 147

Winter Recipes........... 157

Epilogue 186

The Last Word 187

Introductions

Welcome to Oranje Tractor Farm, of which we have custody near Albany, on the pristine south coast of Western Australia. Living this tractor life is possibly as close as it gets, using an oft-quoted saying, to 'living the good life'. We live in the house that we built out of things saved from demolished homes and buildings; bought at auctions and farm sales or salvage yards; or collected from the road-side. It sits cosily embraced by a labyrinth of fruit trees and vegetable gardens which provide sustenance for us, our guests and our visitors; and deciduous trees that provide summer shade on the north facing windows of the house. Whilst giving us a little privacy from the cellar door customers during the busy spring and summer seasons, those same trees are devoid of leaves in the winter, allowing full access to the sun, so that the passive-solar house design works a treat. No air-conditioning is required in our house!

Beyond the sprawling gardens are the vineyards and livestock paddocks, the chooks, the guinea fowl and the sheep. Permaculture principles were at the front of our minds when we designed the garden, or should I say, started planting fruit trees and building garden beds. Design makes it sound like we had a grand plan, when in fact things have just evolved organically – both metaphorically and literally. We have morphed from being a young, energetic couple with 'real jobs' in the health sector whilst building, researching, planting trees and establishing the vineyards on the Oranje Tractor Farm, to creating a successful, albeit tiny, organic wine brand and running a busy cellar door and café, complete with orchard and potager garden. This transformation has taken almost 20 years, so we're no longer young, no longer employed 'in the real world', and the word energetic makes me feel a bit, well, tired. But we are living the dream – our dream and, according to some researchers, the dream that many Australians spend their days pondering. Some, like us, may actually get to plant their vineyard and make fabulous wine. Some – those who perhaps had a crystal ball, and could foresee all the hard work, the economic downturns, and the vagaries of the weather and other aspects of mother nature – might have kept their money simply to enjoy the products made by people like us. Regardless of the pitfalls, however, our lives have been enriched in so many ways that it is difficult to explain in just a few paragraphs. In part this is due to the generosity of our family and friends, who have made very real contributions to our enterprise and to our souls. It is also the result of our involvement in a volunteering scheme called Willing Workers on Organic Farms (WWOOF). Under this scheme we provide accommodation and meals in return for between four and six hours of work per day for each night we host our guests – otherwise known as WWOOFers or woofers. Hundreds of people from across the world have blessed us with their company and assistance through WWOOF over the past 12 years, but the over-riding result cannot be measured merely in the number of hours of physical labour these guests have given us. It is a giant melting pot of friendship and shared memories – memories of satisfying experiences, of recipes shared and meals cooked and enjoyed together, and of projects completed. Our lives have been fundamentally altered as a result of this decade of sharing, in no respect more than in our gastronomic experiences, and we know – from a survey conducted as part of my Gastronomy thesis – that some our former guests have been able to take away many of our unique Australian food and wine customs with them. It is my wish that this publication in some way encompasses this sharing and that it is seen as an acknowledgement to our many former guests, our families and our friends, and the role they have played in enriching our lives. But the real purpose of this book is to allow you, vicariously, to enjoy the experience of meeting some marvellous people, who have engaged with us in nurturing, transforming and developing our piece of the planet; to hear a little about what we have learnt; and to be able to share our eating delights. You are invited to share the joys of *This Tractor Life*.

History

In 1993, we quit our comfortable life in the inner city suburb of West Leederville – in Perth, Western Australia – to pursue a dream. We headed south, way south, to Albany and built a house, planted a vineyard and established a small, organic winery. The wine label was named after the 1964 Fiat tractor that Murray's dad gave us to help with the work, with the (mis-)spelling suggested to us by a German backpacker. That's it, in a nutshell. But, I'll bet you'd like to know more!

So where did this dream materialise from – this dream of owning some land, growing grapes to make wine and feasting on our own home-grown fruit and vegetables? I've often pondered this particular question myself and it wasn't until very recently that the pieces of the puzzle all fell into place. Surfing! Yep, that's what I'm blaming. The call of the ocean, the roar of the sea and the yearning for that last enduring wave... sounds poetic, but what does it have to do with living on the land, making wine and feeding people? Stay with me on this – it's not really all that complicated. We Australians are not only one of the most urbanised populations in the world, but also the most coastal. An extremely large percentage of us live within 50km or 40 minutes of the ocean. Unsurprisingly, given how hot our climate is, our lives revolve around water. And it's not just the beach – rivers, harbours, pools, dams and lakes form the basis of our recreation and often our livelihoods. As a child I lived where the Helena and Swan Rivers met in Guildford, at that time on the urban fringe of Perth, and my best friend Anna and I spent countless hours after school on, in, and around those rivers. Her dad built her a beautiful little wooden boat that we would take on many adventures, most of them picnics on the muddy banks of the meandering Helena River. Our favourite picnic fare was apple fritters, and we would cook them from scratch, on an open fire that we lit ourselves, under the trees that lined the river bank. I'm surprised that we didn't burn the place down – I'm sure we were only aged about 10. Anyway, I digress. Many of our family weekends were spent at my grandparents' beach house, at Waikiki (that's not the Hawaiian variety it's the one south of Fremantle, Western Australia). My father was a very, very keen amateur fisherman, and so was my grand-dad. The beach house was only 300 metres from the calm waters of the aptly named Safety Bay, and thus it was a little piece of paradise. We swam many times each day, returning to the beach house between dips to fuel up from the bottomless supply of sustaining substances my mum kept producing from the little kitchen. We fished and we swam some more. Then, as teenagers, my brothers and I were carried on the wave that became the cult of surfing. It was the cult of the 60s and 70s, where the ultimate dream was to pack yourself, your board and your friends into a VW Kombi van and go in search of waves. Surf movies such as *Endless Summer* and *Morning of the Earth* fed these fantasies, or perhaps drove them? An integral part of this dream was to drop out, find yourself some land (near the best waves, of course) and grow your own vegies. When we finally had access to cars, our partial pursuit of this fantasy involved spending as much time as we could in the Margaret River region – famous now for its world class wines, but only famous then among world class surfers for its fantastic waves! The early pioneers of the wine region had only

recently established their vineyards and were beginning to offer tastings at their rustic cellar doors. We became regular visitors to places such as Gralyn and Pierro, and frequently picnicked under the large peppermint trees at Cullen Wines. Fairly soon, we all included a vineyard into our pipe-dreams! And there, in the pipe, was where I thought it would remain because it became increasingly clear that one needed a small fortune to buy into the wine industry in that region – and there was nowhere else in WA suitable to grow grapes, was there?

Fast-forward a decade or two, and *voila* we had planted a vineyard, built a house and established a large garden. We had created so much work for ourselves that we didn't know what to do. Then, via my favourite radio medium, Our Aunty – or the Australian Broadcasting Commission – I heard about Willing Workers on Organic Farms. You can read all about this fabulous concept in the following pages.

We embarked on being WWOOF hosts with trepidation. After all, inviting strangers into your house before meeting them seems a little scary, doesn't it? None-the-less, in 2001 we accepted a Norwegian couple (and their three kids) as our first woofers. It took me a few days to realise they were just like us: people with dreams, ideas, perspectives and a willingness to share these as well as their labour. Now, some 360 guests later, we can reflect and only imagine what we might have missed out on if we had not taken the risk of participating in the WWOOF scheme.

To the outside world, it would seem that we are living 'The Good Life'. Certainly, thanks to our friends, family, guests and visitors, we have enjoyed a good life. Indeed one of my friends recently said, 'You're living the dream, Pammy'. But let's be clear up front: the good life does not equate to the perfect life, as I'm sure you're already aware, for isn't the quest for a perfect life merely a distraction from reality? And, let's be frank: there's not really a lot that separates a nightmare from a dream, right? It is hard work; we are reliant on weather and the vagaries of nature; and the wine industry in Australia is no longer in its Golden Era. Competition is tough and margins are low, but still we carry on. One of the main reasons is the sense of satisfaction and purpose it provides to us. Our products and services are tangible things – things we have created – and it is extremely rewarding to receive feedback directly from our customers, our critics and our friends about the quality and standard of what we provide.

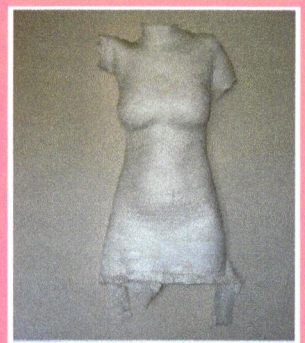

Me

Wine

Ever the scientist, I've been experimenting with food – even the liquid variety – for a very long time. At the ripe old age of 10, one of my best friends (also named Pamela) and I collected some grapes from her backyard and had our first attempt at making wine. It was memorable for all the wrong reasons – it was a messy, sticky job that attracted a lot of bees; it tasted awful; it looked frightening and smelt even worse. I really do not recall what possessed us to perform this alchemy. Although we lived at the door-step of the Swan Valley wine region, our families were not involved in the wine industry and neither parents were big wine drinkers. Sure, our mum's had a sherry at weddings, but beyond a shandy (beer and lemonade for you non-Anglo readers) on a hot summer day, little else alcoholic was consumed by women of that era as a general rule. Men drank beer mostly, although my father liked to buy 'Claret' from the Southern Europeans in the Swan Valley, in one gallon (four litre) flagons. These were the days before the now ubiquitous 'bag-in-a-box' was developed for, ahem, 'commercial' quality wine. Occasionally, my brothers and I would be allowed a sip of dad's precious plonk. I can tell you, it certainly did not have us begging for more. Anyway, until I began my wine science degree in my thirties, I had assumed that while most kids didn't do kooky stuff like make wine when they hadn't even developed a taste for it, anyone who was taking an oenological pathway in life would surely have made a batch or two. Boy was I wrong! Out of a group of about 40 in my wine class at university, only one had attempted wine-making at home, without adult supervision, before the age of 20 – and he was from an Italian wine-making family.

But I digress. My adventures in all things gastronomic began in the kitchen with my mum, an avid amateur cook and later caterer for weddings, parties... anything. Never one to worry much about how much mess was generated in the kitchen, my mother encouraged us all to be involved in cooking. Dad loved to eat novel foods, which was quite unusual for his era and very British background, so mum had licence to be an adventurous cook and early adopter of 'new' cuisines such as Chinese and Italian. Oh how avant-garde, eh? But, do remember that this was the late 60s in Australia I'm talking about!

Although I enjoyed experimenting in the kitchen, as a stubborn, strong-willed teenager I didn't want to embark upon the career as a chef that my mum had held up as the perfect choice for me. By my late teens, I had become intensely interested in the effect of what we consumed had on the human body, and so opted for a nutrition and food science degree. Talk about a pig in mud: not only did we get to study the fascinating fields of biochemistry and human physiology, but also we got to muck about with food in the lab as well as in the kitchen! It was, thus, all a bit disappointing that after completing an equally fascinating post-graduate course and becoming a dietitian, my main role was to encourage people to 'eat less, do more exercise and eat more fruit and veg'.

A few years into this career, following my marriage to Murray, I found myself (with said husband) heading to a friend's wedding in Paris. To ensure that we were up to speed on all things wine prior to our arrival in France, Murray enrolled us in a six-week Wine Education Course. Apart from being great fun, it was life-changing. Not only did we find out that great wines were being made near Albany, Murray's home town, but also that there were two ways of becoming a winemaker. One, which I had already discovered in the middle of my nutrition degree, but ruled out because it would have required moving to Adelaide for three years, was at Roseworthy College, South Australia. The second, we found out, was the wine science degree at Charles Sturt University, and it could be done part-time and externally. Perfect! The rest is history – or is that herstory?

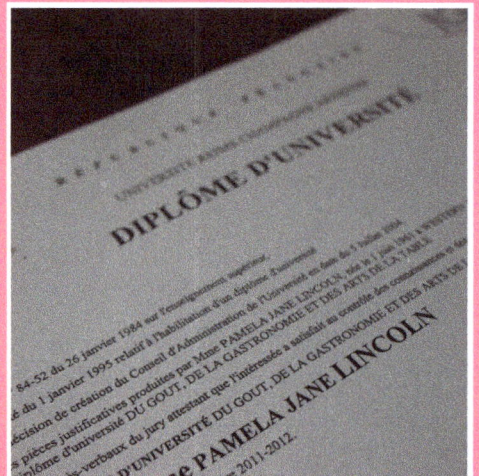 

Food

When we first moved to Albany to pursue our viniferous dream, we derived our income from our professions at the time. It was 1993, and I was fortunate to be able to take up the role of Regional Dietitian with the Health Department. Although it was initially only a 12-month posting, it miraculously became a full-time, permanent position the following year. Serendipitously, the Department's 2 Fruit n 5 Veg campaign had begun in earnest, and I consequently had the need to work with local fruit and vegetable growers to arrange promotions and other events designed to encourage people to eat more fruit and vegetables. It was through this early connection that I, and several of my colleagues at the time, realised two very important things. One was that an incredible diversity of amazing quality fruit and vegetables was produced around Albany. Second was that most of this fabulous produce was trucked all the way to Perth before it was trucked all the way back to the supermarket shelves here in Albany, if it returned at all. There were only two independent green grocers (remember those?) when we first arrived in Albany, and both closed within the next few years. As a public health advocate this was alarming. As a brand new Slow Food member it was concerning. And, as a 'buy local' supporter, it was diabolical. It wasn't until 2000 that, due to the hard work of a woman called Mandy Curnow from the local Department of Agriculture and a band of other forward thinking types, an organisation – to be known as the Great Southern Region Marketing Association – was formed which enabled rectification of these concerns. The association ran a workshop on marketing for local producers, at which Jane Adams – an Australian journalist who'd recently won a scholarship to the USA to study a new phenomenon called 'Farmers' Markets' – was presenting. To cut a long story short, Jane was invited back to run a workshop to help us (Kate, another Health Department colleague; Debra, a fellow Slow Food advocate; some hard-working horticulturalists and farmers and me) begin the establishment of the Albany Farmers' Market. Just after it commenced Murray and I headed to the USA and Europe for my Churchill Fellowship to study organic viticulture and wine making. Perhaps it should have been called '2002, A Food and Wine Odyssey' because we – naturally – explored the many farmers' markets on the west coast of the USA on our road trip from San Francisco to Vancouver via the wine regions. I knew then that the Albany Farmers Market was likely to succeed and be sustained. The public appetite (excuse the pun) for fresh, local produce was huge in the States, and I recognised that, as many trends do, it would be reflected in Australia within a short space of time. This year (2014) marks the 12th anniversary of the Albany Farmers' Market, which has operated every Saturday – rain, hail or shine – since those early days. It gives me such pleasure to be able to source the produce that we are unable to grow ourselves (mostly due to time restraints) for use in the cellar door kitchen on my weekly visit to the market. I acknowledge these hard-working farmers as they are the cornerstone of our local food supply.

Fun

Not one to be left wondering, I've always had a reputation for 'having a go'. It may be the competitive streak in me too, but I never allow my limitations to limit me, and will happily enter a competition if there's a worthwhile prize. *The Churchill Fellowship* is a prime example, as is the *Vin de Champagne* award which gave me (and my friend and fellow winner Debra) the opportunity to spend two glorious weeks in the Champagne region of France in 2004, being wined and dined by some of the many famous Champagne Houses in the land – think Bollinger, Roederer, Pol Roger and more. Many more. This award has been cleverly designed by the French *Comite Interprofessionel du Vin de Champagne* to promote le Champagne (the drink) and la Champagne (the place) and has now been running for 40 years. During this time the consumption of le Champagne has increased exponentially in Australia, so clearly it is working. Recently 45 of the 70 or so previous winners were invited back to Champagne to celebrate the success of the award and were once again lavished with lots of attention, stunning Champagnes and many fine meals. This time we were inaugurated into the very special *L'ordre des Coteaux de Champagne* and congratulated for being masters of Champagne wine. I had to pinch myself to make sure it wasn't a dream. For a wine lover, this must be the best award in the world and it was initiated in Australia. There is only one down side... our friend Terry has nick-named me 'Queen Bubblypiss'. Just don't tell the Champenois, will you?

Him

Meet Mr green-thumb. He's patient, creative and energetic. A man of few words, except when you get him talking about bicycles, wine, gardening, bicycles... and, did I mention bicycles? He's the other half of the Oranje Tractor team and is almost always to be found in the garden. Murray's green thumb was acquired, genetically I think, from his mother Shirley and together they have created the simple yet appealing gardens surrounding the cellar door café. Although I did plant a mighty large number of fruit trees in the early days, about the only role I have in the garden these days is to harvest the produce and to suggest what else might be nice to plant!

Murray's dedication to, and focus on, whatever is important to him are remarkable. He exemplifies sustainable living: recycling and cycling are two major themes that echo through his conversations and, more importantly, through his behaviour. Thanks to Murray, our rubbish bin is never more than half-full, which is quite a feat considering we are generally a household of four as well as being a cellar door and café. The only things that get thrown in the rubbish bin are those that can't be composted, fed to the chooks or recycled. His passion for bicycles and cycling, and all the health, social and economic benefits they can bring to communities and the people who reside in them, is undying. From being the instigator of the local bicycle users' group – which has had enormous success with a series of cycling related projects in Albany over the past decade – to convincing me of the need to produce our 'second label' of wine with a mountain bike theme, the Fat Tyre Wine series, Murray's persistence is a quality to be admired.

Monday night is pizza night here at the Oranje Tractor Farm. For well over 10 years, Murray has created incredibly delicious pizzas, every week almost without exception. Most of our woofers have been indoctrinated with his special techniques – the main one being 'less is more'. Murray insists (and I have to concur) that three is the maximum number of different toppings for maximum flavour. His other secrets to 'the best pizza in Australia' include the fact that pizza bases are always home-made, tomato sauce is rarely used, and that they must be cooked in less than 10 minutes.

Them

Woofers

It is very difficult to imagine our lives without woofers, both the two legged variety and the four legged one. More about the latter, later! Willing Workers on Organic Farms is the organisation from which the acronym WWOOF is derived, and of course woofers are those who sign up to offer their help to people like us who play host. The concept is brilliant, because it's a win-win situation for all. As hosts, we receive valuable labour as well as good company from people who are travelling from far and wide (and even just within this vast land of ours). Woofing is by no means the only form of travel that fosters a non-financial exchange; however, it is one of the few that focuses on agriculture and hence offers a unique opportunity to explore the transference of culture, especially gastronomic culture. It is one of very few opportunities for real agri-tourism or eco-tourism whereby the traveller actually gets to participate in the process of agricultural production. It has also recently been highlighted as a form of *Slow Travel* because it celebrates all that is local, natural, traditional, sensory and gratifying about living in that particular part of the world and, hence, this too is strongly linked with gastronomy.

Our guests, whose ages have ranged from 16 to 72 (we won't mention any names, Ms Kiwi S – but you know who you are!), enjoy a roof over their heads and a belly full of good tucker in return for contributing to our life and our enterprise. The driver for 'volunteering' in this manner seems to be primarily to get out of the tourist traps to meet the locals and to save a few dollars. Increasingly, the desire for a second year of their working holiday is an important motivator for those under 30 years of age to 'go bush' and experience rural life in Australia. A second year working holiday visa is only offered by the Australian government on condition of a three-month stint performing rural work. The benefits to rural Australia cannot, in my mind, be over-stated. Farmers get all-too-scarce labour; families enjoy a great cultural exchange; and rural communities receive great economic inputs. Over the past decade, we have hosted more than 360 individuals in our home from many different parts of the world. Although there have been plenty of the usual suspects – Japanese, Germans, Dutch, French, English and Americans – we've been privileged to also enjoy the company of people from Finland, Taiwan, China, Slovenia, Israel, Iceland, and more besides. Like many hosts in this scheme we treat our guests as part of the family, having them staying under our roof and sharing meals. During their stay, woofers are always encouraged to be involved in meal preparation, in part to share the labour of doing so, but also to enable them to learn from us, for us to learn from them and, importantly, for them to have the possibility of enjoying food with which they are familiar and which they may have missed during their travels. Hence, I have witnessed a wide range of meals from diverse cultures, primarily of the 'home-cooking' style. Given my life-long interest in food, cooking, nutrition and human behaviour, the meals and discussions with the guests have been both enjoyable and fascinating. Along the way, recipes and food-cultures have been recorded and provide a vast repository for investigation, discussion and, ultimately, presentation. Food has been a conduit through which the exchange of experience, knowledge and ideas, caring and friendship has traversed our lives.

Broad generalisations are, generally, politically incorrect. However, in our experience there is considerable truth in some of the cultural stereotypes such as that the Germans are punctual and organised; the French all drink red wine and smoke cigarettes; and the Japanese take lots of photos and are very polite, always. I also reckon that I can pick a French person from 20 paces by the way they cut their salad, or the way they use their hands to express certain words. I can accurately guess who might have set the table for dinner based on the layout of the cutlery, and indeed the choice of spoon. Although not all Germans are tall and blonde, the predominant profile of those who have stayed with us fits the stereotype and they all worked extremely diligently. That said, one of the most meticulously organised people was a French girl, and one of the wildest 'party animals' was a Japanese girl. The funniest guy was a Swede – not typically known for sharing our own sense of humour here in Oz! Martin very quickly developed a rapport and hilarious banter with Murray, who'd taught him a great deal of Aussie slang… although sometimes a little bit would get lost in translation. For example,'Kangaroos in the top paddock' which we had explained as being equivalent to 'having a screw loose' or being a bit crazy, was transformed by Martin into 'a kangaroo loose in the head'. I recently had the opportunity to have a laugh with Martin about it in his own lovely country.

We've also learned that being a younger woofer, say, under 24 years of age, doesn't necessarily equate to being inexperienced or uninterested in interacting with people of a different generation. One of the most delightful guests whose company we've enjoyed was a 19 year old German, Alex. He was not only extremely articulate in English, he was also kind and helpful, intelligent and talented, and of course, hard working. Alex played the piano beautifully, and was planning to study medicine when he returned home. Plus, he liked cooking (and eating, especially cake!!!) and always said how much he enjoyed each meal. Alex also hoped that nearly everything he was served while woofing with us would be in this book. It's people like these who reassure me that there is good in the world, and that the future has plenty of potential. I would not be surprised to hear that he goes on to win a Nobel Prize, such is his talent. Other woofers have gone on to do great things, including – in one case – becoming a professor of biological agriculture at a German university. He – Florian – was also responsible for us spelling Oranje the way we did. Others have married and had kids, and some are in the process of producing the latter.

'Have you had any disaster woofers?' we often get asked – mainly by new woofers! To be frank, only two or three past guests come to mind, but we would happily invite most of our former woofers back to stay with us at any time. What about calamitous occurrences involving woofers? Well, that's a more interesting story… but, no, no snake bites, shark attacks or motor bike accidents. One of the most memorable incidents involved one of our favourite French guests, Jonathan, who was making furniture for us from retired oak barrels. You can read more about Jonathan on page 116.

Workers

We can't really recall how Collyn came to be a part of the Tractor Team, but we're certainly glad he did. Always the gentleman, and ever willing to take on any task, Collyn's presence has become so well established that we'd feel like one of our legs had been cut off if he were to depart Oranje Tractor.

Poppy, our longest serving member of staff, has been part of our lives, almost since we moved here to Albany in 1993. The daughter of dear friends, Erica and Glenn, she is almost family. As a young teenager, she came to work in the kitchen and now she is leaving as a young adult to pursue her life. Her sister Grace has replaced her, and although she is diminutive in stature she is very big on initiative, intelligence and diligence. She's a lot like her dad, Glenn, who has helped us immeasurably over the years doing various tasks with a very practical and under-stated perspective. Wherever we look, there's evidence of either Glenn's hard work or his clever concepts. More recently, we have also secured the services of Sol, the teenage son of other great friends Yann and Louise. Despite his young age, Sol is very adept in the kitchen – hardly surprising given his mother's huge talent in the food-sphere, and her inclusion of all their children in the growing, procuring and preparation of food. We've also enjoyed a few gap-year students and other workers who have made significant contributions to the farm – Robby, Katie and Annie to name but a few. We have been blessed with great help, not just from the woofers.

Friends

Over the years a large number of friends have shared our journey from being public servants to farmers. Many have physically assisted in the many tasks that have been required for us to achieve our goal. Some friends helped us plant wind-break trees onto the bare paddock that we started with in 1995. Others became heavily involved in the laborious work of building our house from recycled materials. Bricks were hand-cleaned of their crumbling old mortar, timber weatherboards were de-nailed and scraped of many years of paint, and large lengths of rough timber were hewn into smooth pieces. Those bits of wood then had to be attached to the building and nailed, the old fashion way – using a hammer! And there was more to come! The vineyard was planted from cuttings taken from other wineries around the region and these had to be cut to length, bundled up and tied before being buried in sand. My friend Judy came all the way from Melbourne to help us do this, and she still talks about 'playing with sticks' today. The planting of those sticks into the ground required a team of friends and family, most of whom were overseen by our friend, and self-designated team leader, Nicole. She prodded and poked us to keep going when the going got tough, and kept us laughing and smiling all the way. There have been teams of grape pickers, comprising our LARD friends (my former university mates) and their children, or other Perth-based friends and their friends plus my nieces Sam and Beck. Many friends have also helped with food and wine events, performing roles from dishwasher, sous chef to 'hydration consultant'. Smaller groups of friends have 'paid' for their accommodation in our guest room by helping out with all many of jobs including disgorging sparkling wine, bottling the Piston Broke Pinot and crushing grapes – often until late into the night. Local friends have likewise helped with a variety of tasks, from feeding the grape pickers to helping out in the cellar door. It is truly remarkable and it is suffice to say we couldn't have done it without the help of such fabulous friends.

Thus it is gastronomy, to tell the truth, which motivates the farmers, vineyardists, fishermen, hunters, and the great family of cooks, no matter under what names, or qualifications they may disguise their part in the preparation of foods.

JEAN-ANTHELEME BRILLAT-SAVARIN (1755-1826) 'THE PHYSIOLOGY OF TASTE' (1825)

His majesty

With his own throne in the cellar door, Merlot the Jack Russell, certainly lives up to the nickname 'His Majesty' that was coined by Collyn (see under Workers on page 29). He doesn't quite lord it over us, but Merlot does have a very good life here on the farm. He's our 'meet n greet' guy, always first to provide a warm reception to visitors arriving at the cellar door. If you're small or if you get down to his level you will also get a wet reception as Merlot loves to splash his tongue around – probably searching for crumbs if the truth be known! He waits patiently for the odd treat from the kitchen, but if none is forthcoming by the end of the day, a sharp yelp reminds me. He has a different version of the yelp for "let me in/let me out" of the door; an excited squeak for "oh my gosh, there's another dog out there... can I go play, can I, can I?"; an agitated bark when he can smell the foxes and kangaroos around; and an almighty ruckus when he's hot on the trail of a rabbit. We occasionally stumble upon a wild bunny on our daily walks around the property. This twice-daily routine is the highlight of Merlot's day, and I can't get away with neglecting my duties because he'll look longingly when it's 'that time', and get decidedly grumpy if I ignore the imploring face for much longer than 15 minutes.

Merlot provides much amusement to us, our woofers and our visitors. His favourite trick is to steal socks, important papers within his reach, and items from open hand-bags in order to attract attention and get you going for a game of chasey. If it is a sock, or sock-like, it will get 'killed' by vigorous shaking; if it's paper (especially loo paper) it gets shredded into confetti-sized pieces; if it's anything else it's usually safe as long as you're chasing him!

We keep a close eye on Merlot, particularly because we lost our first Jack Russell Terrier (called Savvy, short for Sauvignon Blanc) to a tiger-snake bite. It's a reality of life in the Australian countryside that snakes are ever present. They may not always be visible – indeed they tend to slither away silently, long before you get to see them – but they are always around in summer. Sadly, for the snakes, some of them get caught up in the anti-bird netting we place over the vines in January so that the Silver-Eyes and Crows don't eat the whole crop before we harvest in March. This gives us a broad idea of just how many of these snakes there must be in our environs. Tiger snakes are particularly deadly to small creatures, such as dogs and cats, and can kill humans unless urgent medical treatment is given. However, our friend Chris, a local GP, reckons that the only humans that die from snake bite (not the English beverage!) are usually male, drunk and thus do not seek the help they require. Our four legged friends, however, are not usually so lucky due to their small body size. The snake toxin is carried in the lymphatic system, and as our furry friends are small it doesn't take long for the toxin to move to the vital organs where it has its fatal effects. Many a beloved Aussie dog and cat has fallen victim to snake bite. RIP Savvy (the crazy JR) and Bob (Chris's best pussycat).

On a brighter note, a couple of the poor – and very dead – tiger snakes have been used to give Merlot some aversion therapy. I'll spare you the gory details, except to say that any time Merlot attempted to go near the dead snakes, he was subjected to an unpleasant sound. We think (and hope) that he now knows that black slithery things are NOT to be messed with. Some cellar door visitors reported seeing Merlot and a tiger snake heading in opposite directions after almost crossing each other in the garden last summer. Unfortunately, one side effect of the therapy is that we now have a thunder-phobic dog! *C'est la vie!*

Home

Albany is home to Oranje Tractor and has been home to me since 1993. It's a very easy place to live; the climate is good; the environment is superb; the people are friendly; and it has most of the things I enjoyed in Perth – except my old friends – and none of the things I don't like. Albany is also Murray's birthplace, so there is a strong sense of belonging here. It is also a very historic town, as the following anecdote reveals.

In 1788, the British arrived in Botany Bay to proclaim Australia as theirs just days before the French, under the *Comte de la Perouse*, arrived to do the same. Thus, in the race to win Australia it was 'Britain, by a whisker'. Were it not for the inclement weather that prevented la Perouse and his two ships from arriving earlier, Australia would undoubtedly be a French speaking nation today. In fact, in 1772 another Frenchman – Louis de Saint Alouarn – had already landed in Australia (in Western Australia actually) and 'taken possession' for France for King Louis XV. However, the French claim over Western Australia was never secured by a permanent settlement, despite two other French explorers – Nicolas Baudin and Antoine Bruni d'Entrecasteaux – charting the west coast during the closing years of the 18th century. Instead, the British Army Major Edmund Lockyer – who had been dispatched from Sydney, expressly to beat the French to it – established a permanent British settlement in 1826, to be called Frederickstown, later known as Albany.

Albany's history also includes a significant role in the preparation for World War One. A large proportion of the Australian Imperial Forces and the New Zealand Expeditionary Force – later to be known as the ANZACs – sent to Gallipoli to fight for the Allies, last laid eyes on Australian land at Albany. This is because Albany was the gathering point for the 54-strong fleet of ships, carrying 40,000 soldiers and 17,000 horses, which departed for Gallipoli and Europe at the end of 1914. Sadly many thousands never returned. The commemoration of this great loss, through the Dawn Service on ANZAC day, April 25, is reported to have originated in Albany. It is believed that Padre Arthur Ernest White, who had served with the Australian Forces, held a requiem mass for the locally known battle dead at St John's Church in Albany on his return to Australia in 1918. The idea of a dawn ceremony originated from the military routine of waking soldiers before dawn so they could be prepared for attack in the morning's half-light. Today, this special day has become one of Australia's key national days as we acknowledge those who came before us and enabled Australia to develop into the great country that it is.

The natural environment is primarily what brings visitors to Albany, particularly because it is unspoilt and somewhat wild. There are white sandy beaches and bays, often deserted; there are cliffs and rough seas, protected harbours and inlets; there are whales, dolphins and sometimes even sharks. And of course, there are plenty of kangaroos and even some rare, endangered species that have been saved from extinction. As a gateway to the surrounding towns, Albany allows visitors to explore some of the oldest ranges and land formations on the planet (the Stirling Ranges) and some of the tallest trees (the Tree Top Walk near Walpole). On the other hand, wineries, cafes and restaurants mark a change from the rugged natural environment of the wilderness experience. Is it any wonder woofers – and others – say that it is Australia's best kept secret?

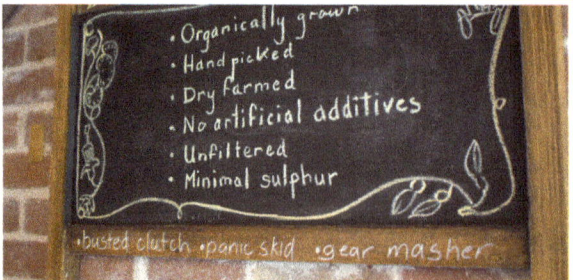

Recipe Journal

Recipe Notes

Measurements are metric and mostly in weights, so make sure you have some kitchen scales. Europeans tend not to use standard cups and spoons in their recipes, mostly weights and milliliter volumes, but I have converted many measurements into standard spoon sizes. One tablespoon is 20mL and 1 teaspoon is 5mL. Unless otherwise stated, flour is plain flour and sugar is regular white sugar. I always use extra virgin olive oil, and believe it is one of the healthiest and tastiest choices, but have used the shorter version olive oil – primarily because I am particularly averse to using the acronym EVOO. The oven temperatures are for a standard oven; if you have a fan forced oven you will probably need to reduce the temperature by 5 to 10 degrees.

I have a Thermomix – the German wonder machine, as we like to call it – so apologies for the frequent mention of its use in these recipes. I've tried to provide recipes that are not reliant on a thermal blender, but I have to admit that I use my wonder machine for nearly everything these days, except of course if I'm slow cooking or pressure cooking. For the former, I use the French Le Creuset pots, and for the latter my German Silit. I love my pressure cooker almost as much as the Thermomix. But fear not, every recipe in this book can be, and has been, cooked without a wonder machine, German or not. Those of you with one should be able to adapt almost all of the recipes easily. If not, email me! There are a couple of soup recipes that require a pressure cooker; they simply will not achieve their wondrous flavours in an ordinary pot – sorry! There's a great website called 'hip pressure cooking' (www.hippressurecooking.com) that provides really clear information about and great recipes for this long neglected piece of kitchen equipment. You'll be amazed just what you can cook in one and how easy – and safe – it is with the modern versions of this often-feared tool.

There are many, many recipes that could have been included here, given that we've had more than 360 people stay with us under the WWOOF scheme. The recipes that appear in this book were selected for ease of preparation, availability of ingredients, interesting cultural background and, above all, their good eating quality. I have attempted to standardise my writing in regard to the method hopefully without losing the unique character of the recipe. Each recipe has been cooked at least twice here at Oranje Tractor – once by the original purveyor of the recipe and again by me or my fellow recipe testers, Max and Antoine. Some have been cooked many more times as they are now some of our favourites. I hope you try many of the recipes and wish you Bon appétit, Itadakimasu, Buon appetito, Bete' avon, Guten Appetit, Dobartek, Eet Smakelijk, Kalióreksi, Smakligmåltig, Bonzer Tucker!

Spring is...

...all about the vineyard

There's nothing quite like the sight of the tiny tips of green, tinged with just a hint of pink, that sprout from the gnarly, grey-brown wood of the previous season's growth to make you give thanks for the splendour of nature. The evidence of new life is palpable when you roam down the rows of grapevines as I do on my daily meanderings with Merlot the fur-child, er, dog. Sometimes it's the critters you notice crawling out of their winter cocoons; other times it's the vegetation springing out of senescence. The focus of our effort and attention from August until almost Christmas time each year is the vineyard, because this is our window of opportunity to ensure the vines have the best start possible. Early attention to the control of the scourge of grapevines – mildew – is the key to a good start for the summer ahead. In early spring this entails regular application of compounds such as copper and sulphur, some of the few permitted under organic accreditation. Murray's dad – Barry – used to keep a tight rein on this until he moved on to that great big green vineyard in the sky. Now we juggle this task ourselves, among all the other tasks on our agenda. Not only do we need to be able to timetable it into our workloads, but the fortnightly spraying needs to be conducted when there's no rain forecast – as the compounds are not systemic and thus wash off in the rain – and when there's not too much wind. A challenge – in Albany – some might say! Then, the ATV (All-terrain vehicle, otherwise known as a quad-bike) needs to be working, as do the sprayer and the dam's water pump. Sometimes, the planets align and all components fall into place. Sometimes, they don't: that's life on the land.

As spring stretches into summer, the vines grow at such a rapid rate that we tend to refer to them as Triffids (those scary plants from the book by John Wyndham called 'The Day of The Triffids'). Within weeks the shoots grow from a few centimetres to almost a metre and by October the tiny flowers have been wind-pollinated and show the first evidence of becoming grape berries. Foul weather at this period, particularly strong winds and heavy or persistent rain, can interfere with this delicate process and result in poor fruit set – meaning a small crop of grapes for the coming vintage. Damp weather and lack of sunshine also encourage the growth of the mildews, so we must do some vine wrestling at this time too. This entails moving the growing shoots into position behind wires on the trellis that supports the vine and its prolific growth during the season, and is more correctly called 'canopy management'. The fifty or more shoots on each vine, making several thousand in total, must be tucked or twisted often more than once during the season. It usually takes at least three passes through the vineyard for this work to be done. If we get side-tracked by other tasks, we might find the shoots entangled with each other due to the incredible tendrils – those curly 'strings' that enable the vines to explore the very tops of the trees in the jungles from which they originated. They are so strong that it takes my full strength to prise them apart, hence the vigorous vine vernacular – 'wrestling'. Another key job at this time of year is breaking off the 'water shoots' – the rampant, random canes that rise from the trunk at ground level. I'm not sure why they're not called 'steroid shoots' instead, given their extraordinary growth rates.

Of course it's not just the vines that are growing at a rapid rate during the warmer months of spring. The grass between the vines, and under the vines, has a tendency to grow at a frighteningly rapid rate as well. Until last year, this required regular mowing between the rows, another time-consuming and fossil fuel depleting duty. Now, we have a small flock of Wiltshire Horn sheep, four legged lawn mowers that I refer to as 'sheep for dummies' because they are relatively low maintenance. Perfect for livestock dummies like ourselves! Despite Murray's upbringing on a dairy farm, and his parents' later switch to pasture-raised cattle farming, he has little knowledge about animal husbandry. And I have even less. These marvellous creatures are self-shedding, so they don't need shearing; they don't need their tails docking; nor do they need crutching or other challenging activities. They just eat the grass and provide manure. Maybe one day they might even provide us with some meat... but neither one of us has the desire to be the death-striker.

Highlights of the season

Vineyard

The onset of spring in Albany is not heralded by any dramatic changes in weather. Rather, with the ever-so-slight increase in daylight hours, comes an ever-so-slight increase in day time temperatures. The spring flush of vegetative growth is very evident on the deciduous trees; less so on the indigenous eucalypts, wattles and sheoaks. The vines also awaken from their winter slumber, having stored the nutrients required for this soon-to-be abundant new growth in their roots at the end of autumn. If we haven't already, we might apply some rock phosphate – a natural form of fertiliser that releases slowly into the soil, without acidifying it and without killing the microscopic world that harbours in the earth. This biomass that includes plant roots as well as bacteria, fungi, worms and other living animals is a significant carbon store – aptly called 'green carbon'. On a global scale, the amount of carbon stored in soils like this is three times as large as the amount of carbon in the atmosphere, that is, the carbon dioxide caused mostly by burning fossil fuels. This is why we are very careful to nurture our soil. We maintain a grassy-weedy cover most of the year and never till or cultivate the soil. We prefer to use foliar sprays of only the nutrients found to be in low supply from our annual soil and leaf testing – all organic, of course – rather than slather the vineyard floor with premixed, water-soluble, possibly toxic, manufactured chemical fertilisers. Over the past 20 years, since we started preparing the land for the house, garden and vineyard, we have been building up the carbon and biomass content of the soil. This has involved composting, mulching and manuring, as well as extensive planting of trees, vines, herbs and flowering plants. We have thus converted a former cow paddock with barely a tree in sight to a flourishing, productive piece of the earth with its own micro-climate, even capable of growing bananas and avocados. The soil carbon level has more than doubled during that time and the number of earthworms has likewise grown remarkably. One of the real benefits of increasing the soil carbon levels is the enhanced water holding capacity, and this is becoming increasingly important with the local effect of climate change being a dramatic reduction in non-winter rainfall. We prefer, for many reasons, to operate the vineyard as 'dry-grown', meaning we don't use supplemental irrigation. This has worked easily for us in the past because we have – historically – had reasonable amounts of rain falling throughout most parts of the year. If it wasn't for the soil carbon improvement, however, the situation would have been quite different given the very dry summers we have experienced in the last few years. We would have found ourselves in quite a predicament because the vines would have needed water to keep the grapes maturing appropriately, but because of the drought conditions our water supply (dam) was at very low levels, so we may not have been able to irrigate even if we had wanted or needed to. Thankfully, those little microbes in the world beneath our feet had done their duty and held fast onto the ample winter rain we experienced. Murray likes to call this 'working with nature instead of fighting against it'.

What shall I tell you, my lady, of the secrets of nature I have learned while cooking?... One can philosophize quite well while preparing supper. I often say, when I have these little thoughts, "Had Aristotle cooked, he would have written a great deal more."

JUANA INES DE LA CRUZ, EPISTOLA A FILOTEA

Woofers

Keiko could cook. A young Japanese woman whose family ran a food business, Keiko was a marvel in the kitchen and was artistically talented too. In this regard she was not unlike most Japanese girls who have stayed with us – they all seemed to be good cooks and clever craftswomen. However, where Keiko streaked ahead of the others was that she did things from scratch. Want tofu? Make it yourself from soy milk. Need soy milk? Make it yourself from soy beans. Right here in our kitchen, using only the beans and a bucket of fresh sea-water, Keiko performed the miracle of transforming dried beans into curd. She made many other Japanese delights, all from scratch and, mostly, they were delicious. And finally, when she departed Keiko not only made the most adorable cookies shaped like us, but a hand-made pop out card featuring the tractor!

Food

Asparagus is the vegetable of the season. I love asparagus. It is one of the three foods that we grow that made me think just recently 'I've made it. I'm in heaven'. For as long as I can remember, the thought of growing raspberries, cherries and asparagus was surreal. Perhaps I'd absorbed much more of my father's British upbringing than I realised. These were some of the foods from his childhood that he would have loved to have grown in his garden, but the baking heat of the Perth summer made it impossible. He had to be content with his tomatoes, beans and silver-beet in that environment.

Ever wondered why we like the foods we do? I've often pondered this, and have come to the conclusion that the human penchant for nostalgia – in all its forms – may be at the heart. It seems to me that we love the foods of our past, particularly those of our childhood, those surrounded by celebration and purpose. These have generational, as well as cultural and geographical aspects. My brothers and I swoon at the thought of freshly cooked Swan River prawns, blue manna crabs and crayfish – foods surrounded by strong memories of fun family traditions, pure pleasure of taste and decadence, respectively. With great frequency during the summer months, mum and dad would pack us all up in the old FJ Holden, complete with prawning net, tilly lamp (AKA gas lamp), buckets and tongs (for the creepy crawly bitey things that inevitably turned up in the prawn net), and head for the banks of the Swan or Canning Rivers near the heart of Perth. The boys would take turns with my father on the other end of the trawling net, and wade chest high into the dark water. It was up to mum and me to wait ashore for the haul to be brought in for sorting. Sometimes the net would be full of clicking, flicking luminescent prawns. At other times, we would have to fight our way through large jellyfish, the occasional cobbler, lots of little fish that we called 'gobble guts' and sometimes large stones, before we could extract the prawns. Luckily, we almost always caught enough for a mini-feast to be eaten later that night. Back at home, with newspaper laid out thick on the kitchen table, mum would pile the freshly boiled prawns and supply us with mountains of fluffy fresh white bread and creamy butter. Nothing tastes as good, to me and my brothers at least!

French paradox

Adeline and Nicholas, a French couple woofing with us recently, cooked many fabulous meals. One that particularly sticks in my memory is a potato concoction that I now call 'potato dauphinoise – Over The Moon style'. It is possibly the most delicious potato dish I've tasted. Anywhere. Ever. *Absolument*. If you know anything about French cooking, you will not be surprised at the quantity of butter, cream and cheese in the recipe. Eating this divine amalgamation of freshly dug spuds and mountains of local dairy produce was an epiphany. Once upon a time, we would have called it 'a heart attack on a plate'. You know lots of that deadly, saturated fat and plenty of salt. The prevailing scientific consensus in the early 1990s was that a high intake of fat, particularly the saturated (animal) variety, was strongly correlated with high rates of cardiovascular disease. This correlation supposedly held true across the majority of western countries, with the marked exception of France. Despite its population consuming relatively high amounts of saturated-fat-laden cheese, butter, cream, etc. the heart attack rate in France was relatively low – hence the paradox. Murray and I learnt about the so-called French Paradox in our Heart Foundation days, and until recently it was simply gospel, irrefutable fact. Later research suggested the protective effect of all that wine that the French consumed was a major confounding factor. Then, it became evident that their vegetable consumption was also protective. Having witnessed the preferred eating habits of the vast majority of our French woofers, I also became aware that they really do eat their greens. I think because we Australians live in an almost perpetual summer, and believe we are the kings of the barbecue, salad aplenty seems logical. The reality is, however, that our salad consumption pales into insignificance when compared with that of our French cousins. Whereas the majority of us here in Oz enjoy salad with a barbecue, the main meal of the day is not typically a barbecue. Numerous pieces of research demonstrate that we Australians are a nation of under-consumers of vegetables, with perhaps one or two vegetables other than potatoes being a typical day's intake. Most of our French guests, however, requested salad at every meal, except for breakfast. In part this may have been due to the availability (from our extensive organic garden), but upon questioning the majority of our woofers indicated that salad was indeed an everyday component of meals in their family lives in France.

Enter Michael Pollan, the phenomenal science journalist with a passion for food, author of *In Defence of Food*, *The Omnivores Dilemma* and much else over the past decade. After initially being affronted by the material in his meticulously researched books (he was throwing all of the gospel and irrefutable facts out of the proverbial window), my scientific backbone was struck by his rational, thought provoking and sound findings. Key amongst Pollan's assertions is that the primary research upon which the cardiovascular disease-saturated fat intake correlation was based is flawed and that fat, especially saturated, is not the bogey. More recently, some medicos and other science writers have taken up the gauntlet and have pointed the finger at sugar and too much carbohydrate-rich food as the main problem with our current eating patterns. My perspective was changing fast, and I continued to desk-top research this thorny issue. There is now good evidence to show that eating less – way less – carbohydrate, and more wholesome animal fat, is not just OK, but it is a healthy way to eat. In fact, hot off the press as I write, the Swedish Council on Health Technology Assessment has published a thorough review of the scientific literature about the most appropriate eating patterns for people with diabetes, and to reverse obesity and related disorders. Its findings, which are usually taken on by the Swedish government as recommendations, are at odds with the rest of the western world's mainstream health authorities in that they are suggesting a low-carbohydrate, higher fat intake. So, to return to the lip-smackingly good spuds cooked by our French friends, it has dawned on me that perhaps it isn't the wine, nor the vegetables (well, salad actually) but the butter and cream that is the protective factor in the French Paradox. Vegetables and butter – what a happy thought! Or maybe it's due to the fact that the French take their food far more seriously. What other country in the world has had its cuisine recognised by UNESCO as Intangible Cultural Heritage? Although the recognition is for

the 'gastronomic meal' (see below), the daily ritual of eating is revered by most French people we've met. From simple table etiquette (which seems to have largely disappeared in Australia) such as waiting for all diners to be seated before starting the meal, and wishing everyone *bon appetit* (or bonza tucker, a term developed by our good friends Rod and Sharon, ironically when we were on holidays together, in France); to always having some cheese with a meal and to lingering and chatting, at length, after the meal, I think that French people really care about their food and about their enjoyment of their meals.

UNESCO Heritage Listing for the French Gastronomic Meal

The gastronomic meal of the French is a customary social practice for celebrating important moments in the lives of individuals and groups, such as births, weddings, birthdays, anniversaries, achievements and reunions. It is a festive meal bringing people together for an occasion to enjoy the art of good eating and drinking. The gastronomic meal emphasizes togetherness, the pleasure of taste, and the balance between human beings and the products of nature. Important elements include the careful selection of dishes from a constantly growing repertoire of recipes; the purchase of good, preferably local products whose flavours go well together; the pairing of food with wine; the setting of a beautiful table; and specific actions during consumption, such as smelling and tasting items at the table. The gastronomic meal should respect a fixed structure, commencing with an apéritif (drinks before the meal) and ending with liqueurs, containing in between at least four successive courses, namely a starter, fish and/or meat with vegetables, cheese and dessert. Individuals called gastronomes that possess deep knowledge of the tradition and preserve its memory watch over the living practice of the rites, thus contributing to their oral and/or written transmission, in particular to younger generations. The gastronomic meal draws circles of family and friends closer together and, more generally, strengthens social ties.

http://www.unesco.org/culture/ich/RL/00437

Spring Recipes

Prawns with garlic and grass fines
Potato dauphinoise
Pastiera napoleatana
Smoked trout whip
Black bean houmous
Okonomiyaki
Pear and passionfruit flan
Choko wedges
Paleo chocolate cake
Mona's backpacker muffins
Cyrielle's killer chicken
Snails
Deep fried brussels sprouts
Asparagus panna cotta

Prawns with garlic and grass fines

We've had some very funny woofers stay with us. Not funny in the weird sense, funny as in hilarious. I totally admire people who can speak another language, and I'm astounded when such people can make jokes or make me laugh. Cyrielle, a French martial arts expert, is possibly the funniest girl ever to have set foot in our house. And she had not just Murray and me amused, but also Jerome – a very funny French guy – and Martha, a 'mature-aged' woofer from the Netherlands. Whether it was pretending to be a masked marauder amongst the vines, or strategically placing corn silk on certain parts of her anatomy, Cyrielle couldn't help horsing around. The Christmas after her stay on the Oranje Tractor Farm, we were thrilled to receive a custom made calendar that she had put together as a memento of her time here, and it bought back such powerful memories of much mirth and merriment. Cyrielle was also a great cook, and made many tasty treats for us to try.

Here's Cyrielle's recipe as she wrote it for us.

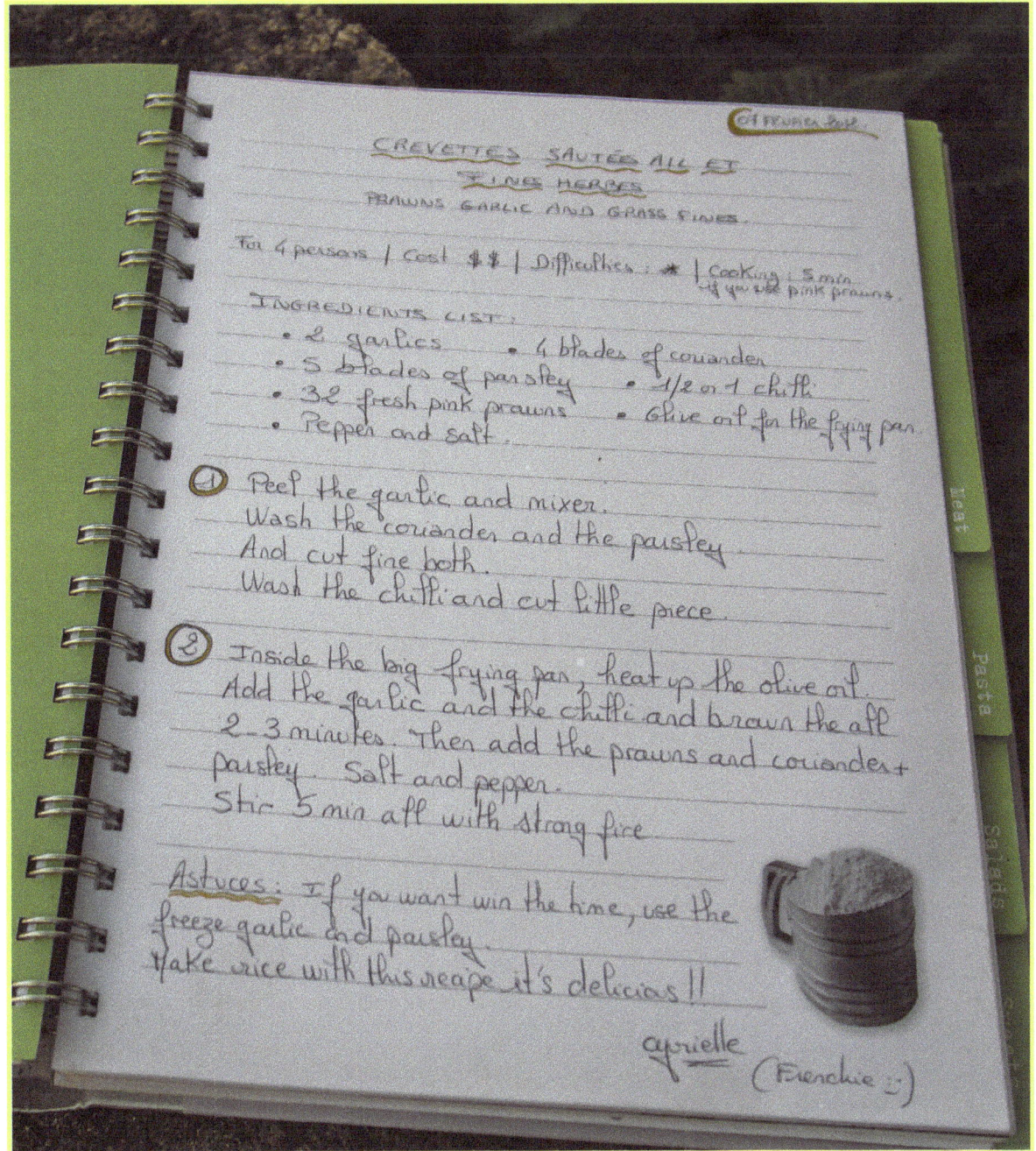

Potato dauphinoise OTM style

Whilst potato gratin is hardly novel, this hearty was elevated to a lofty level by Nico, the French chef woofing with us. It was oozing with deliciousness – and lots of cream too – thanks to the inclusion of our favourite local organic cow's cheese, Over the Moon's Triple Cream Brie. I couldn't resist including the recipe, just as I couldn't resist having a second serving.

This strictly is not the classic dauphinoise because cheese is not 'permitted' in the traditional version – but many cooks, even French ones, include a little. You could modify it more by adding thyme leaves if you like, or even use a combination of sweet potatoes with regular ones. However, I have to say, why mess with perfection?

SERVES 6
Time 40 minutes

1kg waxy potatoes
400mL milk
Pinch of salt
30g butter
2 cloves of garlic, crushed
Pinch of salt
Pinch of grated nutmeg
300mL cream
150 to 200g Brie – Over the Moon Triple Cream if you're lucky enough

1. Set oven to 180°C
2. Wash and dry the potatoes, then peel and slice them into thin, even slices 3-4mm wide. Do NOT rinse potatoes but DO pat them dry with kitchen towel.
3. Put the milk into a pan, add potatoes and salt. Bring to a gentle boil for 10 minutes, stirring occasionally. The potatoes should not be soft at this point. Strain the potatoes carefully, reserving some of the milk in case your dish needs topping up after the cream is added.
4. Grease a casserole dish with the butter, then scatter the garlic over the base of the dish. Tip the potato slices in gently. Sprinkle salt and nutmeg over the slices.
5. Pour on the cream and arrange slices of Brie over the top of the potato. Top with a little of the reserved milk if the potatoes are not fully immersed in cream.
6. Bake for 20 minutes, until the top is lightly brown and bubbling. Remove from oven and allow to sit for 10 to 15 minutes to enable the rich, velvety texture to develop.

This pairs well with roasted meat as a side dish, or can be served with a green salad as the main course of a lovely vegetarian lunch.

Pastiera napoleatana Italian Easter cake

This quintessential southern Italian cake was made for us by Daniela, an Italian girl, but it wasn't at Easter time. There are, apparently, several versions of this but we were assured that this was the 'traditional one'. Good job there were no other Italians around at the time, as there would no doubt be a debate about that! My research indicates that wheat berries are typically used for this cake and so is orange blossom water. However, I love the simplicity of Daniela's version which uses ingredients most of us would have on hand. Of course, if you have a surplus of milk, as we often do you can enhance it somewhat by making your own ricotta. It's really, really easy – just Google it, but make sure you add a splash of cream to the method for extra flavour and texture.

SERVES 8
Time 2 hours

3 lemons
175g butter
1 teaspoon cinnamon
Pinch of salt
150g Arborio or Carnaroli rice
500mL milk
350g flour
1 teaspoon vanilla essence
375g sugar
5 whole eggs and 3 egg whites
2 oranges
500g ricotta cheese

1. Set oven to 180°C.
2. Grease a high-sided 20cm round cake tin.
3. Grate the rind of the lemons. Mix one third of this rind with 25g of the butter, $\frac{1}{2}$ teaspoon of the cinnamon, salt, rice and milk in a pot. Simmer for 25 minutes, then let it cool in the pot.
4. Mix the flour with 175g of the sugar, the remaining 150g butter, 1 egg and 2 of the egg whites in a bowl. Stir well to create a soft dough then cover the mixture and allow to rest in fridge for 30 minutes.
5. Grate the rind of the oranges and mix it with the ricotta, the rest of the sugar (200g) and the remaining lemon rind, vanilla essence, $\frac{1}{2}$ teaspoon cinnamon, 4 eggs and the remaining egg white in a large bowl. Pour the cooled rice mixture into the ricotta mixture and stir to combine.
6. Remove dough from the fridge and set aside one third of the dough for the lid. Roll the remainder out large enough to line the base and sides of the cake tin. Fill the lined cake tin with the rice-ricotta mixture. Use the remaining dough to make a lid to put on top of the cake tin. Bake 1 hour and 15 minutes.

Smoked trout whip

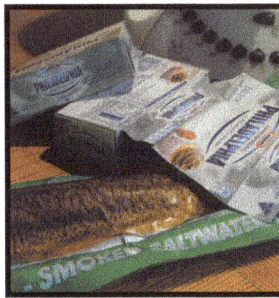

This is a cellar door favourite. If we take this off the platters the locals and regulars are very disappointed. And, all our woofing guests – except the strict vegetarians – have loved this too. It is deceptively simple to make, and the stunning flavour of the local smoked trout shines through. The trout was the brainchild of Tony Smith – who is not only one of the region's pioneering vignerons here in the late 1960s – but also responsible for developing this high quality product. After 'retiring' from Plantagenet Wines (the winery he established in the early 1970s), Tony kept himself busy by raising Rainbow Trout in his and neighbouring farm dams that were too salty for stock or for irrigation, thus putting an otherwise under-utilised resource to good use. Gotta love that? He then tinkered with a few recipes for seasoning and smoking them, and with the help of the local butcher Les from Plantagenet Meats in Mount Barker, came up with a mouth-watering final product – Bouverie Trout. You might find it hard to get, unless you're down this way, but well worth it, I can assure you (if you haven't been fortunate enough to taste it here at Oranje Tractor Farm).

SERVES 8
Time 15 minutes

1 whole smoked trout
300g Philadelphia cream cheese
1 tablespoon plain yoghurt

1. De-bone the trout and put the flesh into a blender.
2. Add the cream cheese and yoghurt and blitz until smooth.

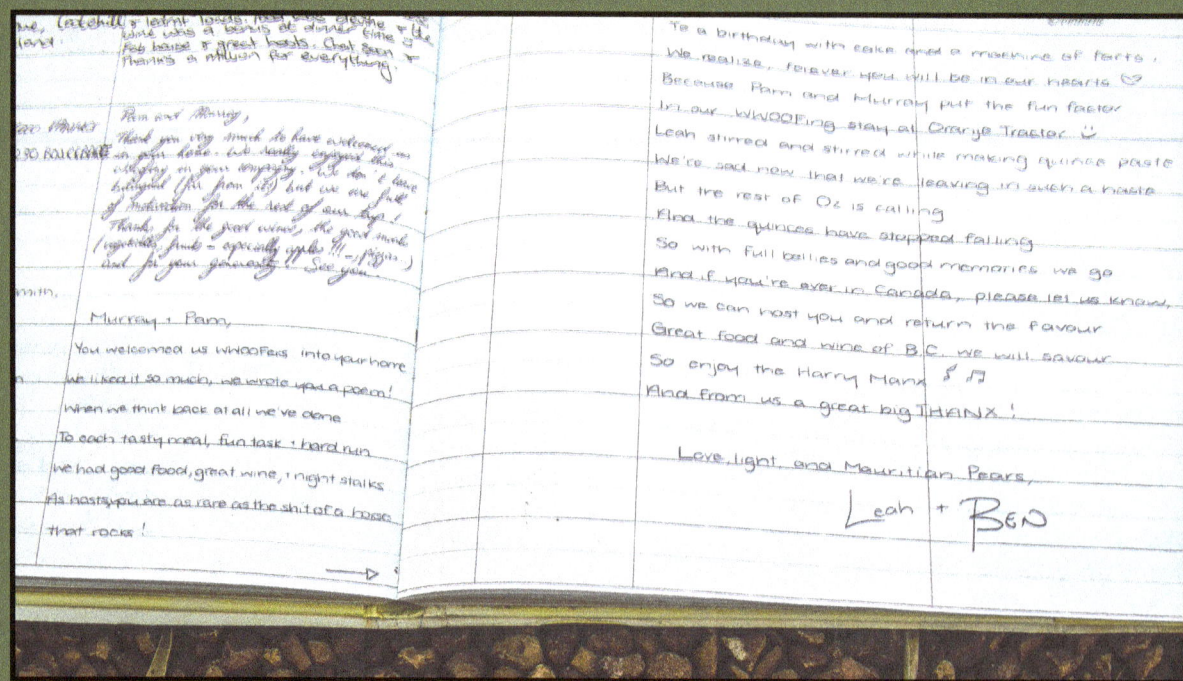

Black bean houmous

We thought we had tasted every dip known to humanity. That was until Ben and Leah, a most gorgeous couple from Canada, introduced us to Black Bean Dip. We felt a strong kinship with this couple as they were both planning to study in the health field when they returned to Canada. They subsequently did, and they got married. I would have loved to go to the wedding, but the vineyard and cellar door responsibilities precluded this. Oh, and that stuff that doesn't grow on trees.

Ben or Leah, or both of them, once worked at a Tex-Mex place, so this recipe originated somewhere in the land of the maple tree. This version uses dried beans, mainly because we can't readily get canned black beans here. If you can, good luck to you – it will save you a bit of time. However, this dip is worth the effort even if you can't.

SERVES 6 – 8
Time 45 minutes

2 cups black beans
Pinch bicarbonate of soda
Olive oil
1 clove garlic, minced
2 tablespoons Tahini
Juice of 1 lime
$\frac{1}{8}$ teaspoon chilli powder
1 sprig coriander, chopped
Salt to taste

1. Soak the dried beans for at least 4 hours in cold water.
2. Discard the water and put beans into a large pot or pressure cooker and cover by about 10cm with fresh water. Do NOT add salt at this time as it will stop the beans from softening, but DO add a small pinch of bicarbonate of soda to retain the colour, and a teaspoon of olive oil.
3. Cook as per directions on pack, around 40 minutes in a pot or 10 minutes in a pressure cooker, or until beans have softened. If you forgot to soak them, just cook for longer (double the time).
4. Cool, strain and place in a blender with the garlic, Tahini, lime, chilli and about 1 tablespoon of olive oil.
5. Process until smooth, adding more oil if desired.
6. Add a pinch of salt or more to your taste and the coriander and mix by hand to incorporate.

Enjoy with vegetable crudités, crackers or corn chips.

Okonomiyaki

Despite having had dozens and dozens of Japanese woofers over the years, it wasn't until a year ago that we were introduced to this easy, one pan dish meal. Kenji, a French woofer, made a return visit to us with his Japanese girlfriend. Despite her difficulties with speaking English, she was a welcome guest, especially after she whipped up some of these pancakes for us. I now use a mandolin to do the slicing of the vegetables, although I'm somewhat reluctant to recommend these very sharp kitchen tools as a general rule. Perhaps our nickname for it – the thumb-slicer – might give you a clue as to why? Be afraid, be very afraid when you use a mandolin, and always use the guard.

Time 20 minutes

For each person:
1 cup finely sliced vegetables
– carrot, cabbage, onion are a good base
to which you could add red capsicum and mushrooms
1 tablespoon olive oil
2 eggs
30g flour
50mL water or cold stock, preferably Bonito fish stock
1 Nori (sushi roll seaweed) sheet
HP sauce
Mayonnaise – Japanese Cewpie brand is traditional

1. In an oven-proof frying pan, heat the oil until hot and sauté the vegetables for a few minutes.
2. Turn the heat down, then put a lid on the pan and allow the vegetables to sweat a little until *al dente*. Turn off heat and allow to cool a little.
3. Meanwhile, crisp the seaweed sheets by waving them over the gas flame of your stove or similar, using metal tongs, until they will crumble when crushed.
4. Put the eggs, flour and stock or water into a bowl and whisk until smooth.
5. Pour this onto the vegetables and place on stove over moderate heat until the egg has almost set.
6. Finish off by placing under a hot grill for a few minutes. Remove and squirt the HP sauce in a zigzag pattern over the pancake, then do the same with the mayonnaise but going in the perpendicular direction.
7. Garnish with crumbled seaweed and serve.

Pear and passionfruit flan

This recipe was devised to use fruit that we frequently tire of eating fresh, despite how delicious they are separately or combined. Sometimes we host woofers who just can't get enough passionfruit and will happily devour dozens per day. One in particular springs to mind, but no names, no pack-drill but this Fräuleinwill know who she is! But we always seem to have passionfruit garnishing the kitchen table most times of the year. Luckily we only have two passionfruit vines. This is not the case for pears. Murray decided – just as he did for the blood plum trees – that if one is good then 10 is better. I often wish he'd asked for my advice in that department particularly when the table (and fridge, freezer and pantry) is groaning with more pears than seems humanly possible to eat. It's a very quick and easy flan to whip up, particularly if you have a freezer full of pre-poached pears. Dealing with our fruit-gluts is one of the main tasks for our woofers when they're sick of doing vineyard work. I don't know of a woofer yet who has left our place without having sliced, diced, chopped, poached, stewed or otherwise dealt with one type of fruit or other. This recipe is also gluten-free so has become a favourite of late. Replacement of the sugar with honey or natural sugar substitutes, and the butter with coconut oil, will make it paleo-primal friendly.

SERVES 6
Time up to 1 hour

4 pears
1 cup sugar
1 cup water
150g butter
180g caster sugar
4 eggs
200g ground almonds
½ teaspoon baking powder
Pulp from 4 passionfruit or ¾ cup of berries of choice

1. Peel and cut pears into halves, quarters or eighths before poaching in a light syrup comprising 1 cup sugar and 3 cups water, with or without a vanilla bean or cinnamon stick. If you are short of time and don't have any poached pears in the freezer, you can use canned sliced pears.

2. Brush the inside of a flan tin with melted butter. Set oven to 180°C.

3. If you have a Thermomix or similar, you can simply put all of the ingredients except for the fruit in the blender and mix on speed 4 until well combined. If not, use a mixer or regular blender cream the butter and sugar, then add the eggs one at a time, mixing well between each addition to form a smooth batter. Fold in the ground almonds and baking powder.

4. Pour the batter into the prepared tin. Place the sliced pears on top of the mixture in a pattern of your choice. Press the slices into the mixture slightly, drizzle with passionfruit or berries and sprinkle with a tablespoon of extra sugar.

5. Bake for 30 minutes or until the mixture is set.

choko wedges

One choko vine produces many chokoes. Too many, actually, for one family, even with a constant stream of guests, so I was always looking for new ideas for them. That was until a kind customer at the cellar door said that she had a stash of recipes for them because she had chokoes growing too. And a stash it was! A big fat envelope arrived a few weeks after the lady's visit, full of pages of hand-written recipes for chokoes. I'll admit to not having tried them all, but this one is great. They really are a versatile, if not particularly strong tasting, vegetable – strictly, they are a fruit – and can be used as one might use zucchini, adding bulk to curries and stews, or cubed and pan-fried in butter. They can also be poached in light syrup and eaten as a dessert. There's an urban myth that says that a certain fast-food chain uses chokoes in place of apples in their apple pies. What's not mythical is that people of my parents' generation say that their parents often used chokoes in place of pears during the Depression years.

SERVES 4
Time 20 minutes

2 chokoes, peeled and sliced into sticks or wedges
$\frac{3}{4}$ cup self-raising flour
3 tablespoons ground rice
Pinch of salt
$\frac{1}{4}$ teaspoon turmeric
$\frac{1}{2}$ cup water
Oil for frying

1. Heat enough oil to come about 5cm up the side of a large pot to 190°C.
2. Make a batter by combining the flour, ground rice, turmeric and salt in a bowl.
3. Add the water and mix until smooth.
4. Dip the choko pieces in the batter and fry until golden.
5. Drain on kitchen towel.

Serve as is or with a cream sauce.

CHOCOLATE
+
ALMONDS
„PALEO"

Paleo chocolate cake

Now a cellar door favourite, this is a guilt-free chocolate indulgence. It is a modification of an Italian chocolate torte that is traditionally made with hazelnuts, sugar and a lot more eggs. Luigi and Beatrice, two of our relatively few woofers from Italy, shared the recipe with us. Beatrice's family were winemakers in the Cupramontana region near the north-east coast of Italy and we enjoyed a fine bottle of their Verdicchio – a white wine – they had brought with them. If I can get fresh hazelnuts from Pemberton (a side crop from the new truffle industry that has popped up there) I will use those, or a mix of almonds and hazelnuts. Quite frankly most of the hazelnuts you find in supermarkets are old and have lost their flavour, so almonds (Australian ones) are the next best thing.

This decadent chocolate cake is gluten free and is made with no added sugar, so meets the criteria for Primal eating. If you substitute coconut oil for butter it is then suitable for a Paleo diet. But don't let that put you off, as this nut cake is delicious in its own right and, if you have a Thermomix, ridiculously easy to make. If you haven't, don't worry, it's not that hard with a regular blender or food processor. You can make a chocolate ganache or simply dust it with icing sugar and serve with double cream. This recipe makes a small cake because it is quite rich and you will not want large serves. However, it works well if you double all the quantities to make a large cake if you're entertaining a crowd.

SERVES 6
Time 1 hour

200g dark chocolate, broken into pieces.
Lindt eating chocolate is great because it is thin and thus easy to break and to process. You can often find it at reduced prices, so buy up big then – as long as you can resist the temptation to eat it in the meantime!
200g whole almonds
100g dates, seeds removed
1 tablespoon cocoa powder
150g butter at room temperature
4 large eggs
2 teaspoons baking powder (gluten free)

1. Set oven to 160°C
2. Grease a 22cm round cake tin or line with baking paper.
3. If using a Thermomix, put chocolate pieces, nuts and dates (check each one to ensure no seeds are included) into the bowl and blitz on speed 6 for 10 seconds, then increase to speed 8 for another 5 to 10 seconds to get a uniform, crumbled mixture. Add the cocoa, butter, eggs and baking powder and mix for 20 seconds on speed 5.
4. If using a blender, process the chocolate, almond and dates separately and place in a large bowl. Put the butter and eggs into the blender and process until well combined. Add this to the chocolate mixture along with the cocoa and baking powder. Stir well with a wooden spoon.
5. Tip the cake batter into the tin and place in the oven for 40 minutes. Check to see if cake is cooked with a clean skewer – it should come out clean. If not cooked, leave for another 5 to 10 minutes and check again. Remove from oven and allow to cool for 15 minutes before turning out onto a rack to cool completely.

Mona's backpacker muffins

Mona was a good friend of Sabine, our potager garden creator, and also from Germany. When she made her 'backpacker' muffins – single serve, ready in 3 minutes from go to whoa – I thought they must have been a German invention. But no, she picked up this handy little recipe during her travels in Australia. I'm surprised that only one of our woofers shared a microwave muffin recipe with us, because 'fast' food, like instant noodles, is very popular when backpackers are on the road. I modified Mona's recipe to make it gluten free, and it works a treat. Perfect for those moments you need something sweet for yourself or a friend and there's no time to spare.

SERVES 1
Time 3 minutes

3 tablespoons (50g) ground almonds
2 tablespoons cocoa powder
1 tablespoon honey or sugar substitute such as stevia
1 teaspoon butter or coconut oil
½ teaspoon vanilla essence
Pinch of salt
1 egg

1. In a large microwave-proof mug combine all the ingredients.
2. Stir well.
3. Microwave on high for 1½ minutes. If your microwave is not a 750 watt version, you may need to cook 10 seconds longer.
4. You can spice it up if you like by adding cinnamon before cooking, or even a hint of chilli.

Eat while warm.

Cyrielle's Killer chicken

This simple dish is surprisingly tasty, although I do think we are blessed to be able to buy the sweetest, tastiest carrots locally as well as the fabulous free-range Mount Barker chicken. Cyrielle – who signed herself as Cyrielle Killer in our guest book after Murray explained to her that he used Serial Killer as a memory jogger to help him remember her name (after 360 or more guests, it is a struggle sometimes) – cooked this for us one very memorable night. We ate it with a salad and some corn on the cob, quite a novelty for Europeans who typically view corn on the cob, like pumpkin, as animal food although they are catching on, it seems.

SERVES 4
Time 45 minutes

4 chicken breasts, skin on
4 carrots, peeled and cut into small rounds
1 tablespoon olive oil
2 teaspoons cumin seeds
4 sprigs of coriander
1 lemon, zest and juice
Salt
White pepper

1. Dry the chicken with paper towel and season the skin with salt and pepper.
2. Heat the oil in a frying pan and place chicken skin side down.
3. Cook until golden, then turn the heat down and add the carrots and cumin.
4. Stir carrots frequently and flip chicken over to cook other side. Add more oil if needed.
5. Once the chicken and carrots are cooked, drizzle with lemon juice and zest and season with salt and pepper.

Serve with a green salad.

Snails

Several ex-woofers, all French of course, prepared snails for us to eat using our own home-grown garden snails. Snails are very popular in France and have been eaten since Roman times. In the south of France they are typically eaten barbecued at family gatherings with plenty of garlic, salt and butter. Elsewhere in France snails are cooked outside of their shell then returned for service. As an example, in the Alsace region the snails are simmered in Riesling then returned to the shell and served with some garlic and herb butter. The common brown snail is edible and often in abundance around here at Oranje Tractor Farm.

To prepare the snails one must purge them of their stomach contents, so they are usually put into a wooden box and fasted for 5 days. Most of our snail-preparers allowed the snails a diet of fresh herbs such as rosemary or thyme, reportedly to give more flavour. After this period the flesh is removed from the shell, rinsed in fresh water and seasoned well with salt and pepper. The flesh is then returned to the shell and the snails cooked on the barbecue, flesh side up, for 5 to 10 minutes, and finished with a dob of butter and a sprinkle of parsley.

Matthieu and Marie, who were from Perpignan, cooked our snails on the barbecue with plenty of butter and salt and, I have to say, the little morsels were tender and tasty. This lovely couple were on their way back home to get married, and our place was their last port of call in Australia. Matthieu was a talented chap – an engineer – even though he had received local fame in South Australia for cooking snails because the local ABC radio ran a story about him. 'A French chef', they called him. You can see it for yourself if you Google his name, Matthieu Duflot, and ABC radio. He thought it was amusing and we all laughed at yet another lost-in-translation moment.

Matthieu told us that in the south of France it is expected that each person will eat 20 snails at a barbecue. I was roundly applauded for meeting this target. Although I ate them with gusto, we haven't bothered to repeat the exercise ourselves. But I included the method just in case you wanted to try it yourself. It's one way of getting rid of the little blighters.

Deep fried brussels sprouts

Brussels sprouts typically elicit a strong response from almost everyone, don't they? You either love them or hate them depending mostly, I think, on how you were first served them as a kid. If you were presented with crunchy, lightly cooked sprouts you probably are in the former camp. Sadly, many people only remember the slimy, overcooked, smelly little bitter balls of their childhood. Part of the problem too is that some of us are genetically, more able to detect bitterness when we are children, so there's a double whammy. However, I implore you to try this recipe – just once – even if you've sworn off Brussels Sprouts for life. Or better yet, get someone else to do it for you without fore-warning, so your preconceptions are not primed! There is truly something magical about this method of cooking these mini-cabbages. I've cooked Brussels in every way possible and, apart from boiling them to death, most methods are enjoyable. But this method, which does not actually leave them tasting oily at all, results in a fantastic eating experience. The main reason that this recipe is included is because Alex, one our favourite gastronomically-inclined woofers begged me to include it in the book after tasting them. Enough said?

SERVES 6
20 minutes

500g Brussels sprouts
1 – 2L peanut oil or similar high smoke point oil
1 lemon, juiced
1 tablespoon butter
salt

1. Trim the sprouts of any damaged or tough outer leaves. Cut in half lengthwise. Remove 2 or 3 outer leaves on each half and place into a colander. Pour boiling water over these leaves, then rinse under cold water and set aside. They should retain a lovely bright green colour and an al dente crispness.
2. Rinse the halved Brussels and dry completely on paper towel.
3. Heat the oil in a deep pot and using an appropriate thermometer (eg., a Candy thermometer) to 190°C.
4. Carefully place about 1 cup full of sprouts into the oil and fry for about 4 minutes, until the leaves start to brown. Remove with a slotted spoon and drain well on kitchen towel. Keep warm by placing in an oven set low (100°C) while you cook the remaining halves. Ensure the oil returns to 190°C before adding the next batch of sprouts.
5. Meanwhile, heat the butter in a frying pan over medium heat and add the blanched leaves to heat them through.
6. Add the deep fried sprouts and season with salt and lemon juice.

Serve immediately. For a more exotic effect, you can sprinkle with some sesame oil and a splash of fish sauce, and maybe even some sliced chillies.

Asparagus panna cotta

This savoury panna cotta makes use of one of my favourite spring vegetables, asparagus. We have our own small asparagus patch which I guard carefully, although we rarely get a whole meal's worth of spears at any one time. Instead I usually eat them one by one on my morning walk with Merlot the dog. Philip and Shelagh Marshall's Asparagus – some of the key producers involved in the establishment of the Albany Farmers' Market – is my favourite source of fat, fresh asparagus from late August until Christmas. The flavour is so good that I usually just pan-fry the spears lightly in butter and olive oil and serve them with a Hollandaise sauce. I don't bother peeling the ends of the spears as they are rarely stringy. Antoine, one of our Michelin-starred French chefs, cooked this divine dish using the first of Marshalls' 2014 crop of asparagus for a Champagne dinner we held recently, and garnished it with shaved truffle. The purpose of the dinner was to celebrate the truffle season and to help my friend Louise train for the Vin de Champagne Award. I kid you not, it was fantastic.

MAKES 10 SMALL SERVES
Time 10 minutes preparation and cooking, 2 hours chilling
1L cream
20 asparagus spears
3 shallots, peeled and diced
7 gelatine leaves, placed in a bowl of cold water to soften
Salt to taste, Olive Oil

1. Trim the top 5cm off each asparagus spear and set aside. Heat frying pan, add oil and fry diced shallots until soft. Add asparagus stems and toss a few minutes. Finally add the cream and simmer gently until asparagus is cooked but not too soft.
2. Add the softened gelatine leaves and stir well before taking off the heat. Allow to cool a little, blend the entire mixture and then sieve it to remove fibres.
3. Pour the liquid into 10 small oiled Dariole moulds. Allow to cool almost completely then place on a tray in the fridge for at least 1 hour. They will keep for a couple of days covered in the fridge. If you do keep them in the fridge overnight or longer, remove them 1 hour prior to serving to allow the gelatine matrix to soften a little.
4. In the meantime, blanch the asparagus tips in boiling salted water for a few minutes, then drain and place immediately into iced water.
5. Serve these as a garnish for the panna cotta, which you will need to turn out onto a serving plate to un-mould.
6. You can do as Antoine did and repeat the above process, with prosciutto (10 slices per litre of cream) instead of asparagus, and make a two-tone panna cotta.

Don't forget to shave the truffle over the top!

Summer

Is the time for... Events

As the weather warms up, and the vines put on vast amounts of growth, the number of cultural, sporting and other activities increases around the region. There is also an influx of visitors to the cellar door, especially around Christmas time. Thus, any events we hold are planned for either before the summer holiday period or after.

Our first major event involving a celebrity chef (as much as I detest that term, it's difficult to find a more appropriate descriptor) was a long table lunch as part of one of the first Great Southern Taste festivals. Russell Blaikie, a well-known Western Australian chef and a truly lovely bloke, was engaged by the coordinator of the festival to serve up a fabulous feast of fresh local food, and we provided the venue, the wine and the staff. Naturally, we asked our good friends to help out. Little did we realise that Russell required every bit of refrigeration space we had and more. Little did we realise also, just how many plates and glasses, and how much cutlery 75 people would require. And, little did we realize that the temperature would, uncharacteristically, rise to a sultry 35 degrees C out there in the garden. The last required some drastic measures, coordinated by Mark, our 'hydration consultant'. Nicole, with her naturally effervescent and assertive style became the maître d'; Christine was the sensational sous chef; Therese and Rex were exemplary waitpersons-cum-dish pigs; and Murray and I did our best to keep the flow of water and Oranje Tractor wine trickling into the thirsty mouths of our guests. For a bunch of amateurs, we did well. Despite the lack of sufficient facilities, particularly refrigeration, Russell did exceptionally well. There were no hitches, no dramas and no knives thrown, not even by me! Everybody seemed to be enjoying the day, and many said how much fun they had or how much they couldn't wait until next year's event. Russell subtly suggested that perhaps some additional refrigeration might be wise for a future feast. Gotcha, Rusty.

Maggie Beer has been, for a very long time, an inspiration to me. I know I'm not alone in that department for she is one of Australia's best loved cooks. She's humble, she's creative, and she's self-taught and clever. It is no wonder she received the Australian (Senior) of the Year Award a few years ago. Can you imagine, then, the state of excitement I was in after I was asked if I would agree to interview her as part of the Albany Writer's Festival in 2009, sort of Parkinson-style or Andrew Denton-esque. And, if that wasn't enough to make me spin on the spot, the festival coordinator asked if we would host a literary lunch at Oranje Tractor with Maggie as the guest speaker. Is the Pope a Catholic? As the photos depict, it was a huge success despite a little rain that coincided with Maggie's walk through the garden with Murray and a troupe of hardy souls.

In another fabulous coup we hosted a literary lunch for Emmanuel Mollois, the cheeky French chef who regularly appeared on Poh's Kitchen – a cooking series on Australia's ABC television. We helped him launch his cookbook, with my friend Louise being the cook and patisserie... with much accolade from both Monsier Mollois and the happy guests!

Eat Food, Not Much, Mostly Plants

MICHAEL POLLAN

Highlights of the season

Vineyard

The vines have grown so rampantly by this time of year that we often struggle to keep them under control. The weekly tucking, the fortnightly spraying and the frequent slashing of the grass occupy us until Christmas time. In between we'll have been on the look-out for beasties: snails, grasshoppers and weevils, all of which love to feast on the foliage. Being organic means we can't use toxic chemicals to blast such pests away. Instead, we carefully monitor to determine the pressure the vines are facing from these and other critters. Often, the bugs are just at nuisance level so we simply tolerate them. If their numbers are rising, we or the woofers might hand collect them and either turn them into eggs – by feeding them to the chooks – or create something exotic like escargots! Sometimes there's a plague and there's not much that can be done... but thankfully that's only happened a couple of times. Typically, it has been grasshoppers or locusts – and they eat anything green in their path. The resulting defoliation of the vine affects the grapes because without young, metabolically active leaves there is no photosynthesis to supply the energy for ripening and flavour development. By late January it is time to get the nets out, as the next and biggest pest is flocking in for a feast. Little birds called Silver-eyes do a large amount of damage in the vineyard. They peck at individual berries on the bunches of grapes, opening them up to attack by mould and other diseases, and potentially turning the whole bunch 'sour' as fermentation may commence prematurely. Then there's the Australian Raven or what is commonly called the black crow. A majestic looking bird, with luminescent black plumage, this is one of the smartest avian species. Fortunately, they eat whole berries so don't wreck whole bunches of grapes. None-the-less they will, if left to their own devices, gobble up the whole crop. The 28 Parrots, or Port Lincoln Ring-Necks are the next in line for a feast on the vines... hence the need for anti-bird netting. The netting goes over the vine rows and is designed to exclude even the little Silver-eyes. Using a contraption that is reminiscent of a basketball hoop, which our friend Glenn welded for us to go on the back of the ATV trailer he built, the net (which sits on the deck of the trailer) is passed through the hoop and then out over the vines. It is a labour-intensive task, requiring at least three people: one to drive the ATV; two to stretch the net over the vine row. What follows is even more arduous in terms of time: pulling the net so that it reaches the ground, and fixing any holes larger than a 10 cent piece because those darn little Silver-eyes can slip through very easily. For some people, this is a mind-numbing job; for others it is a joy as they are doing something productive, and in the great outdoors.

 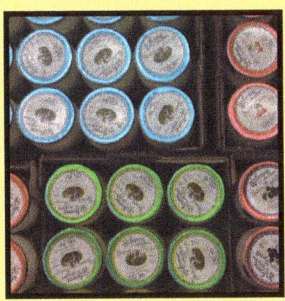  

Woofers

Some of our woofing guests stay for months, most for a few weeks and others are with us for only a short time. But the duration of their stay does not necessarily reflect either our memory of them or their impact upon our lives. One example is that of Priscilla, a young Canadian woman who stayed with us for a few weeks to learn as much as she could about wine in a practical sense to supplement her already expansive wine knowledge. During her stay she shared with me many stories about the amazing things she learned through a Gastronomy Diploma in France she had completed, so her stay was fruitful on both sides of the woofing gate. I was hooked immediately and began to investigate the possibility of undertaking the program myself. In late 2011, after receiving a small sum of money from my deceased Pop's estate, I flew off to Paris to do the intensive two-week component of the program. It truly was amazing and I'm so grateful to Priscilla for introducing me to the course. Each day's lectures, given by eminent researchers and professionals, were supplemented with incredible food experiences varying from visiting the huge 234 hectare Paris market (where the public is not generally permitted) at three o'clock one morning, to dining in Michelin-starred restaurants, and from having a whole day's lesson at the Cordon Bleu cooking school, to enjoying amazing themed dinners such as an authentic Medieval banquet.

As part of the requirements for the diploma, I had to undertake a thesis and it became obvious that our gastronomic cultural exchange through woofing would be an ideal topic. What I found, after surveying previous guests of Oranje Tractor Farm, was not especially surprising to me. However it may surprise some – who perceive Australia to be a gastronomic cultural desert – to learn that 78 percent of the respondents said that they discovered new food and drink experiences that were different from their own whilst travelling in Australia, and that the majority of those respondents reported adopting these during their stay. Very few reported adopting what they perceived as negative habits, but of those who did, eating 'fish and chips' and 'toasted bread and sandwiches' were the main ones mentioned. Similarly, woofing was reported by three quarters of respondents as being 'a valuable cultural exchange, particularly in relation to gastronomy'. They enjoyed the obvious and ubiquitous barbecues, kangaroo steaks and Pavlova with the various WWOOF hosts with whom they had stayed. They also valued the relaxed way of Australian life and eating, and the fact that they experienced 'real gastronomy – not a tourism product'. What did surprise me a little was that many woofers, particularly Europeans, were themselves surprised at the diversity and freshness of fruit and vegetables available in Australia, and at the quality of the wine! This is reflected in some national tourism research conducted recently which showed that whilst Australia's food and wine offering was only considered the sixth most important single reason for visiting Australia by those surveyed who had not visited, it was considered the number 1 destination for food and wine by those who had visited Australia.

Food

Perfectly ripe peaches, flavoursome juicy apricots and truly tasty tomatoes are just some of the joys that summer in our garden brings. In this neck of the woods, zucchini, corn and red capsicum are late summer additions that reek of long, languid days and somewhat – this is Albany, after all – balmy nights. Prolific and plentiful, the produce from the garden quickly overwhelms the kitchen table at this time of the year. There are never enough hungry mouths to feed, nor enough time to preserve all of this goodness for the days and months to come. It is a big garden!

Any discussion of those summery vegetables that we associate with Italian cuisine sends me into somersaults, and not just because of the delicious meals created from tomatoes, zucchini, corn and capsicum. As a nation, we Australians have been guilty of what's called the 'cultural cringe' – a feeling of embarrassment about perceived lack of culture. I think most of us (and the rest of the world) now realise that we do, in fact, have our own culture and sense of identity in Australia. Sure, it's not the same as the culture of 'the mother country' – long seen as the UK — but it is a set of beliefs, values and behaviours that separates us from people in other countries. This extends to our gastronomic culture too, but there remains a considerable degree of hesitancy to take pride in our diverse and delicious food habits and history. We seem reluctant to lay claim to food traditions, habits and customs that may be up to 200 years old. But, consider this. Before Columbus sailed the ocean blue, Italians did not have capsicum, corn, courgettes or tomatoes in their culinary stew. They ate what most of middle Europe subsisted on: alliums such as onions, leek and garlic; cruciferous vegetables such as cabbage and cauliflower; and roots such as carrots and turnips. There were also no beans or eggplant until the Americas was conquered. Also consider that restaurants have only been a part of French gastronomic culture since the mid-1770s – and the French were the first in Europe to move from a very private dining experience to eating in public. That wasn't really that long ago, yet we are quick to see tomato-rich spaghetti Bolognese, eggplant Parmigiana and corn polenta as 'traditional food' of Italy.

101 ways to wash the dishes other trivial tit-bits

Whilst on the topic of gastronomy, it would be remiss of me if I didn't point out that there's more than one way to wash the dishes. Trust me, I reckon I've seen it all and I believe it is a reflection on the realities of life. In countries such as Australia and Israel, where water shortages have always been a risk if not a reality, our use of the precious liquid is thrifty. One of the very first waves of woofers we hosted were Israelis, and they all washed the dishes with the smallest amount of water possible. Rather than half fill the kitchen sink, as we normally do, the dirty dishes are all scrubbed with a soapy sponge before being quickly rinsed. Contrast this with most of our European guests who, if left to their own devices, would have the tap running the whole time the entire meal's worth of dishes are moistened, scrubbed and rinsed individually. Given that we here at the Oranje Tractor Farm are reliant on rainwater for all our supply, this practice is quickly curtailed and we ask them to proceed using the mostly standard method. This entails half-filling the sink, as per usual, to be followed by a rinse sink. Many woofers expressed their preference of this over the other common method they experience in Australia of simply washing in soapy sink water, followed by drying with a tea-towel. There seems to be considerable angst among our European cousins about detergent residue, and perhaps they have a point? But they do have plenty of water in Europe that they can use. The Japanese fall mid-way between the two extremes of thrift and thoroughness of rinsing, in that they pre-scrub the dishes like the Israeli method, but rinse well under running water – a bit like the way they bathe. The English, surprisingly, all want to use a plastic tub inside the sink, and perhaps this reflects a time when nothing was wasted and the water used to irrigate the garden?

Nothing has surprised me more, however, than the apparent disappearance of a spoon specifically for the consumption of soup at the European table. I have questioned family and friends, colleagues and acquaintances alike here in Australia, and all report not only having soup spoons in their cutlery drawer but also to prefer using them over a table or dessert spoon for consuming soup. A soup spoon in Australia (and in Britain) is round, with a concave bowl. A dessert spoon on the other hand is egg-like in shape. Both have roughly the same capacity. The rounded bowl is ideal for sipping hot soup from the side of the spoon, whereas the dessert spoon has been designed so that the first third or there-about can fit into the average mouth. Whenever our European guests set the table, we invariably end up with soup spoons to eat our breakfast cereal and our dessert with, and sometimes we get a dessert spoon to use with the soup. It drives me a little bit crazy, I have to admit. The German, French and Italian guests we have hosted have all reported that they only have the one type of spoon in their home, which is usually the same shape as our table or dessert spoon. It is unclear when the round soup spoon disappeared from European flatware, but is certainly seems to have been commonplace in the 18th century. I looked in a department store in Paris recently, and sure enough noted only dessert spoons as part of cutlery sets. Ah, the wonders of having a wide variety of house guests: how else would I have ever known that marvellous bit of gastronomic trivia, eh?

Summer
Recipes

- 87 Blaukrant
- 89 Jeff's chilli sauce
- 90 Greek-style stuffed summer vegies
- 91 Lemon & rosemary cake
- 93 Crème citron
- 95 Susan's famous 'best ever' cheesecake
- 97 Scottish mess
- 99 Mirror cake
- 101 Pflammenkuchen
- 103 Pepernoten
- 105 Finnish farmer biscuits
- 107 Max's grandma's french fries
- 109 Pancake cake
- 110 Avocado ~~coffee~~ smoothie

Blaukraut

Nearly every year since we started hosting people under the WWOOF scheme we have had guests with us at Christmas. Some hosts choose not to take woofers at this busy time of year, but we enjoy the warm fuzzy feeling we get from sharing our celebrations with these Christmas orphans. It is the first time most woofers have had a warm Christmas or a beach one. What seems so ordinary to us can be a kaleidoscope of wonderful scenes to others. Kathi, Jan and Max, gap-year students from Germany, were our most recent Christmas woofers. These charming, smart young things were thrilled to be invited to our beach breakfast on Christmas morning with our friends Tracey and Ed (and many more of their friends).

Having eaten sauerkraut in Germany and Austria over recent years I was very glad when Jan gave made this slightly different, but equally tasty, version of kraut, using red cabbage, which turns blue under the influence of the acidity: hence "blau". This is one to eat and enjoy rather than store before consuming which one does with sauerkraut.

SERVES 4 TO 6
Time 90 minutes

1kg purple cabbage
50g butter
2 tablespoons brown sugar
1 apple
1 onion
4 cloves
2 bay leaves
3 juniper berries
100g red wine
100g apple juice (unfiltered or cloudy juice is best)
Pinch of salt
Splash of red wine vinegar

1. Remove outer leaves of the cabbage and discard. Slice it in half and place cut side down for stability before slicing finely. Dice the apple and onion.
2. Heat the butter in a large pot until melted, then add sugar and cook until caramelised.
3. Add the apple and onion and cook gently until the onion is translucent – do not brown it. Add the cabbage and stir. Place a lid on the pot and allow the cabbage to sweat for a few minutes.
4. Add the wine, apple juice, vinegar, spices and salt and simmer for one hour. It may be necessary to add another dose of the liquid. Cook until liquid disappears, then adjust the flavour with salt and a little more vinegar if necessary.

This goes particularly well with mashed potato and black pudding (blutwurst in German, boudin noir in French). In Cologne I was introduced to eating typical brewhouse cuisine by ex-woofer Alex and his mum. The meal I chose was blutwurst and sauerkraut, plus Himmel und Erde (Heaven and Earth) which comprises mashed potato with apple sauce swirled through it. The whole thing is served hot and was very delicious. And so was the accompanying Kolsch beer – served cold, I might add.

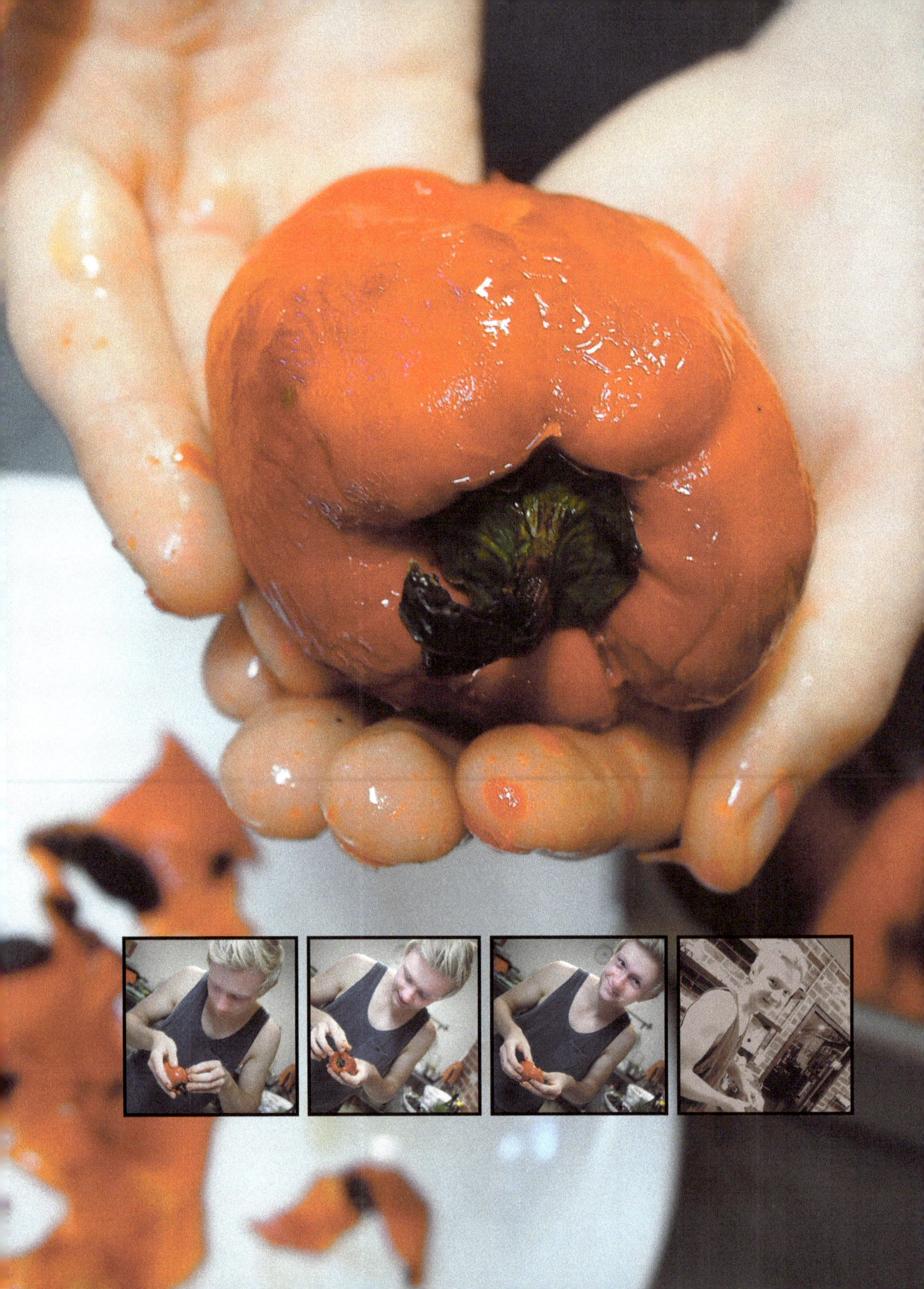

Jeff's chilli sauce

Jean-François – otherwise known as Canadian Jeff – was our first French-Canadian, and we couldn't have asked for a better representative. Jeff had chosen to woof with us because he and his brother were planning to establish a vineyard on their father's farm near Quebec. It may seem bizarre to think of grapevines growing in the snow-laden fields of Canada, but they do grow well and, in certain areas, produce amazing wines such as Ice Wine. This superbly sweet, viscous and concentrated dessert wine is luscious and world class. It's similar to the botrytised wines of France (think Sauternes) or De Bortoli's Noble One. Anyway, Jeff was a delightful guest and made a return visit with his lovely girlfriend Karin a few months later, after they had spent a couple of months cycling around New Zealand – all of New Zealand. We fully expected to receive a wedding invitation from Jeff and Karin the following year, but sadly that wasn't to be the case. Jeff did, however, set up his vineyard. He also sent his charming brother and his girlfriend to visit us. We can't wait to taste the first wine.

Unlike their French cousins, French-Canadians do eat and enjoy hot and spicy food. Jeff was particularly partial to fiery feasts, and created this sauce from our limited selection of chillies. It's hot, but it is mighty addictive and was enjoyed by Jeff on almost anything. You can try it with couscous, roasted potatoes, scrambled egg or even as a dip – if you're brave! Alex, one of our fabulous German woofers, also loved this sauce on everything. He helped me make a batch so he could eat it with gay abandon.

SERVES MANY
Time 30 minutes
20 fresh red chillies, small Serranos or similar
1 large red capsicum
2 cloves garlic, crushed
1 teaspoon cumin seeds
1 teaspoon coriander seeds
1 teaspoon caraway seeds
50 to 100mL olive oil
1 teaspoon or more of salt, to taste

1. Set oven to 180°C
2. Prepare the chillies using gloves: chop stems off, slice lengthwise and remove seeds.
3. Put the capsicum on a baking tray and place in oven for 15 minutes or until the skin starts to blister or split.
4. Dry roast the spices in a frying pan until fragrant but not brown. Allow to cool then crush finely in a mortar and pestle or spice grinder.
5. Remove the capsicum from the oven and, using tongs, place in a plastic bag for 10 minutes. Remove carefully as there will be hot liquid inside. Strip the skin from the capsicum and discard it, along with the seeds, liquid and white pith.
6. Put the capsicum, spices, chillies and crushed garlic into a blender and blitz for as long as it takes to get a smooth paste – 30 seconds in the Thermomix!
7. Add the oil slowly while continuing to mix until you have a pourable sauce, then salt to taste.

Store in a glass jar in the fridge, covered.

Greek-style stuffed summer vegies

We've been lucky enough to have enjoyed the company of several chefs during our WWOOF hosting decade. Mostly, they were French and male - ah yes, poor us! Natalie, however, was our first and only Greek girl and chef. Accompanied by her very funny boyfriend Nikko, Natalie shared much of her Greek culture and cooking with us and a German couple, Steffi and Olaf. Natalie also decided she'd take the opportunity of being so far away from home to do something radical. One night, after preparing these delicious stuffed vegetables, she disappeared with Nikko and the hair clippers. Shortly after, they reappeared, with Natalie sporting a number 1 haircut. That's short, very short. Poor Steffi burst into tears at the sight of Natalie's almost bald head. The rest of us, unfortunately, just laughed even harder! The Greeks eat more vegetables than their other European counterparts, apparently due to the Greek Orthodox religion which requires regular fasting, especially avoiding animal flesh. Hence the Greeks have created a myriad of wonderful vegetable dishes, including this one. Olive oil and vegetables are the basis of the healthy Mediterranean diet and, when fresh, seasonal produce is selected, are tasty enough to be a meal in their own right. You can, however, add minced meat to the stuffing for these vegetables for a variation. Use a good quality extra virgin olive oil for this dish as it is an important part of the flavour.

SERVES 6
Time 60 minutes

6 medium sized capsicums, any colour
6 large tomatoes
1 medium zucchini
2 tablespoons olive oil
1 onion
2 cloves garlic

100g long grain white rice
1 teaspoon dried mint or several sprigs of fresh mint
1 tablespoon pine nuts (optional)
Small whole potatoes

1. Set oven to 180°C.
2. Cut the tops off the tomatoes and capsicum, reserving them as lids. Scoop out the seeds and white pith of the capsicum and discard. Scoop out the seeds and flesh of the tomatoes and place in a bowl.
3. Peel and dice the onion, and dice the zucchini. Heat the oil in a pan and gently fry the onion until translucent, then add the garlic, zucchini and rice. Stir to coat the rice in the oil mixture then add the mint leaves and sauté for 2 or 3 minutes. Tip in the tomato flesh and a tablespoon of water and cook again for 2 minutes. Season with salt and pepper to your taste and turn off heat.
4. Oil a casserole dish large enough to fit all of the tomatoes and capsicums in. Fill the vegetables with the rice mix and put the lids on. Pour enough water to come 2cm up the side of the vegetables.
5. Place small potatoes in between the tomatoes and capsicums to fill the spaces, season with salt and pepper and add a tablespoon or two of olive oil.
6. Cover with foil and cook in the oven for 40 minutes. Spoon some of the liquid over the vegetables then serve.

Delicious with feta (our favourites are Albany's Over the Moon or Ringwould feta) and some crusty bread.

Lemon & rosemary cake

This cake is simply one that was created because of an abundance of both lemons and rosemary in our garden. It is a modified version of one that contains breadcrumbs, but we like to offer our cellar door guests a gluten free option, so I removed the bread from the original. You can add it back in if you prefer a fluffier cake: replace 50g of ground almonds with three quarters of a cup of wholemeal breadcrumbs. Most of our woofers are enthralled to see lemons growing on a tree. I guess most citrus trees don't like frost, let alone snow, so it would be fairly challenging to grow your own in continental Europe.

SERVES 6 to 8
Time 1 hour

Cake
200g almonds
180g sugar
160mL olive oil
4 eggs
Grated zest of 2 lemons
1 teaspoon rosemary leaves, finely chopped
1 teaspoon baking powder (gluten free)

Syrup
Juice of 2 lemons
90g sugar
100mL water
2 sprigs of rosemary

1. Set oven to 180°C.
2. Prepare a cake tin by lining with non-stick baking paper or oiling.
3. Coarsely grind or blend the almonds. You don't want them to be like almond meal, you want more of a crumb consistency.
4. In a bowl or blender combine the almonds, sugar, oil, eggs, lemon juice and rosemary. Mix in the baking powder and pour the mixture into the tin.
5. Bake for 35 minutes. Test cake to see if it is cooked using a clean skewer – it should come out clean. If it needs more cooking, turn the oven down to 160°C and cook another 10 minutes. Remove from oven and allow cake to cool in tin for an hour.
6. To make the syrup, put all of the ingredients into a pot and simmer until the mixture has thickened a little, but not coloured. Remove the rosemary and discard.
7. Once the cake is cool, turn it out onto a serving plate. Use a skewer to prick about 20 fine holes in the cake, and then pour the syrup over it slowly, a spoonful at a time to allow it to soak in.

Garnish with a rosemary sprig or two and serve with fresh double cream.

Crème citron

Sometimes simplicity is sublime. One of Murray's favourite sayings is 'less is more', particularly in relation to pizzas. I think if you have fresh, local and seasonal produce – preferably organic – then his maxim can be applied more generally. This simple dessert epitomises this idea and was introduced to us by Adeline, a French girl. It would be as appropriate for a family dinner as for a special dinner party. You can easily double the recipe and it will serve 8 or more if you reduce the size of the ramekins you use.

SERVES 4
Time 35 minutes

6 eggs
170g caster sugar
2 teaspoons grated lemon rind
100mL lemon juice
150mL cream

1. Set oven to 160°C.
2. Beat the eggs with the sugar until creamy.
3. Add the lemon rind and mix again.
4. Slowly incorporate the juice while mixing.
5. Add the cream and stir until combined.
6. Pour the mixture into 4 oven-proof ramekins and place these into a high-sided baking tray half filled with hot water.
7. Cook for 20 minutes until just set.

Susan's famous 'best ever' cheesecake

If we awarded medals to our woofers then Kiwi Susan, originally from the USA, would get one for having the best age to work output ratio! Susan was not old in either her outlook or her enthusiasm. Nor did she look her 'three score years and ten'! A remarkable woman, with incredible tenacity and *joie de vivre*, Susan stayed longer than most and entertained us with stories from her past as well as blitzing the garden like she was a spring chicken. It was truly sad to see her leave, but we think we might see her and her flaming red hair again down this neck of the woods, as she fell in love with Albany and plans to move as soon as she can! Susan's famous (well, that's what she told us) all American Cheesecake is a classic, and very, very moreish.

SERVES 6
Time 45 minutes

Base
1 packet of Arnotts Granita (or similar) biscuits, crumbled
100g butter, melted

Filling
500g Philadelphia cream cheese, softened
2 eggs
100g sugar
1 teaspoon vanilla essence

Topping
400g sour cream
50g sugar
1 teaspoon vanilla essence
1 teaspoon ground nutmeg

1. Oil a 22cm spring-form pan and set oven to 180°C.
2. Combine the biscuit crumbs and melted butter, then press onto the base of the pan.
3. Make the filling by beating the cheese, eggs, sugar and vanilla together until smooth. Pour this onto the base and place in the oven to cook until set, approximately 25 minutes. It may darken in colour, but don't worry. Remove the pan from the oven and allow to cool.
4. Turn oven up to 200°C.
5. Mix the sour cream, sugar and vanilla essence until combined then pour this over the filling.
6. Bake for 10 minutes, then remove from oven and allow to cool.
7. Sprinkle with ground nutmeg and chill a few hours before serving.

Scottish mess

Based on the English dessert Eton Mess, this simple yet stunning-looking creation was made for us by one of our few Scottish woofers, Colin. He and his friend Andy were woofing along with a French girl, Lauranne; an Austrian girl, Elisabeth; an English lad, Dominic; a German girl, Christina and her funny Spanish boyfriend, Ruben. It was like a United Nations gathering, but I suspect much more fun. It was around vintage time, so there was plenty of work to keep everyone busy. We ate a wide variety of cuisines and had many, many laughs together. This dessert was just one of the many lovely meals I didn't have to cook. The crew made some amazing pizzas and you can see some of them on page 117. More correctly known as Cranachan, this divine mish-mash of fresh raspberries, cream, meringue, oats and whisky is served in a tall glass in order to appreciate the full visual beauty of the contrast of colours and textures. It can be a very quick dessert to make if you have some pre-made meringues available; otherwise, get cracking and start whipping those egg whites.

SERVES 5
Time 35 minutes

50g caster sugar
80g rolled oats
200g raspberries
250mL whipping cream
10mL whisky
1 tablespoon honey
4 small meringues

1. Set oven to 180°C.
2. Mix the oats with the sugar and spread out on a baking tray. Put the tray into the oven and toast for 20 minutes or so until golden. Remove and allow to cool completely. Colin soaked the oats overnight in whisky before toasting them, but this is not essential. Although I have to say it was delicious!
3. Break the meringues into small chunks.
4. Warm the honey to make it runny and then add the whisky.
5. Whip the cream until soft peaks form. Reserve 8 or so raspberries, then gently add the rest to the cream with the whisky and honey, mixing until well combined. Finally, add the cooled oats to the cream and stir until just combined.
6. Spoon into tall glasses and garnish with crumbled meringues and reserved raspberries.

If you love raspberries as much as I do, you will be going back for more.

Mirror cake

This cake was quite the production. Cyrielle, our funny French woofer, took most of the day to create this cake and the end result was fantastic. Mirror cakes are popular in France and may be made of a variety of fruit.

SERVES 6 to 8
Time 90 minutes

Base
3 eggs
3 tablespoons sugar
80g flour
20g cornflour

Mousse Filling
300g raspberries, puréed
300mL whipping cream
½ cup sugar
6 gelatine leaves or 6 teaspoons gelatine powder

Mirror Topping
250g (approximately) raspberry purée
200mL apple jelly (or use 50g sugar and 150mL water)
2 teaspoons or 2 leaves of gelatine

1. Set oven to 180°C.
2. Prepare a 22cm spring-form pan by greasing it with butter or oil. Make the base by beating the eggs and sugar until pale and creamy, then fold in the flours. Pour into the prepared tin and bake for 10-15 minutes, until lightly browned. Remove from oven and allow to cool.
3. If using leaf gelatine, place the 6 leaves into a bowl of cold water, ensuring there is enough water to fully cover them. Allow to sit in the bowl until the leaves soften, approximately 10 minutes. When you are almost ready to use them, remove from the bowl and place into a cup. Pour a small amount (about ½ a cup) of very hot water into the cup to dissolve the gelatine and stir well. Allow to cool for 5 minutes.
4. If using powdered gelatine, sprinkle the 6 teaspoons evenly over a small, heat proof bowl of water (containing no more than half a cup of water). Do not stir. Allow it to 'bloom' for 10 minutes.
5. Press the puréed raspberries through a sieve to remove the seeds. Heat the purée slightly and add sugar, stirring until it dissolves, then stir in the gelatine and turn off the heat. Mix until the gelatine is dissolved and the liquid is smooth. Whip the cream until soft peaks form, then stir in the cooled but not cold berry-gelatine mix. Pour this onto the base of the spring-form pan and place into the fridge or freezer. Allow to chill and set firm.
6. To make the topping, sieve enough puréed raspberries to get 4 tablespoons of seed-free liquid. Put it in a small pot with the 2 teaspoons or 2 leaves of gelatine and the apple jelly. Heat gently and stir until dissolved. Turn off heat and allow to cool a little. Pour the topping over the filling and return to the fridge to chill and set completely.

Slice. Eat. Enjoy.

Pflammenkuchen

German blue plum cake was first introduced to us by Steffi and Olaf. Steffi had a fine repertoire of traditional German meals, which we enjoyed often as they stayed with us some months and revisited us a few years later with their first-born child. They now have two, but I have yet to meet the second as we were not able to rendezvous on my last visit to Europe. They had hoped to immigrate to Australia, and even attained eligibility for a sponsored-work visa but weren't able to find an employer. Anyone want a hard working Master Landscape Gardener?

This traditional cake is synonymous with autumn as it is reliant on fresh plums. Frozen ones are simply too runny. We have used many varieties of plum (remember, we have 10 plum trees, and some are not blood plums), but the best are either the said blood plums or the prune or sugar plum. The latter are drier but incredibly sweet, and because they're not as delicious fresh as the other varieties, this is a great use for them – particularly as I soon get sick of drying them. On the other hand, the gorgeous colour of the blood plums makes up for their messiness. One can use yeast dough for the base or, for the time poor, this short crust style one is just as traditional. The resulting cake can be served warm for dessert with a big dollop of double cream or for afternoon tea. Heck, I've even eaten it for breakfast with some plain yoghurt. It's not overly sweet like some cakes so it seems light.

SERVES 8
Time 50 minutes

Dough
400g flour
2 teaspoons baking powder
Pinch of salt
150g sugar
125g butter
3 eggs
4 tablespoons cold milk

Fruit layer
1.5kg plums (any will do, but sugar plums or blood plums are best)
5 tablespoons sugar combined with
2 teaspoons ground cinnamon or more

1. Set oven to 200°C.
2. Wash the plums and halve or quarter them, remove and discard the stones.
3. Prepare a deep-sided, large, rectangular baking dish by greasing well with softened butter.
4. Make the dough by first mixing the flour, salt and baking powder in a bowl. In another bowl, cream the butter and sugar until pale and creamy then add eggs one by one, beating after each addition.
5. Fold in the flour gently then pour the milk in and mix until combined. You should have a soft, pliable dough that you now spread on the prepared dish. Working swiftly, place the plums so that they cover as much of the dough as possible. Place the dish into the hot oven and cook for 30 minutes. Check that the base is cooked by inserting a clean skewer. If it comes out clean it is cooked. If not, cook another 10 minutes and check again. Once cooked, remove from oven and cool for a few minutes before sprinkling the cinnamon-sugar mix over the plums.

Serve warm and enjoy fresh; it will not be as good the next day – if there's any left!

Pepernoten

These spiced biscuits were introduced to us by Helena, a recent visitor from the Netherlands. Pepernoten are typically eaten on 5th December for Sinterklaas or Saint Nicholas' Eve, which is the traditional day of gift giving for the Dutch, akin to our Christmas Day. Saint Nicholas is the patron saint of children and of sailors. As well as the exchange of gifts in the evening, children leave their shoes by the fire with the expectation that in the morning they will find that Saint Nick has left Pepernoten and a host of other tasty treats in the shoes.

SERVES MANY
Time 30 minutes

First make a Speculaaskruiden by combining the following:
15g (or 2 tablespoons) ground cinnamon
1 teaspoon ground cloves
¾ teaspoon ground nutmeg
½ teaspoon ground white pepper
½ teaspoon ground dried ginger
¼ teaspoon ground cardamom

Batter
150g butter
125g brown sugar
2 teaspoons Speculaaskruiden
250g self-raising flour
Pinch of salt
4 tablespoons milk

1. Set oven to 160°C and prepare 2 trays with non-stick baking paper.
2. If using a food processor, pulse the butter and dry ingredients together until a crumbly texture is achieved. Add the milk gradually to form a firm dough – you may not need all the milk. Roll the dough into small balls, place on the tray and flatten slightly with a fork or a flat bottomed glass.
3. Bake for 15 minutes.

Finnish farmer biscuits

We haven't had too many people from Finland stay with us. This may be because there are not really a lot of them, or maybe they don't woof? Anyway, the one and only Finn to apply to us for a woof post was a delightful young woman, so we assume that the rest of them will be good folk too. Maybe one day we will visit to find out – these biscuits are enough to entice me.

SERVES MANY
Time 50 minutes

250g butter
190g sugar
45g golden syrup
1 egg
¼ teaspoon vanilla essence
400g flour
1 teaspoon baking powder
80g roasted almonds, chopped finely or ground coarsely

1. Cream the butter and sugar in a bowl. Add the rest of the ingredients and mix until a firm dough is formed. Wrap in cling film and refrigerate for 30 minutes.
2. Set oven to 180°C.
3. Cut the dough into 4 pieces and roll each one out to form a log about the diameter of a 50 cent coin. Slice each log into 0.5cm disks and place on a baking tray lined with non-stick baking paper. Repeat until finished.
4. Put tray into oven and bake 7-10 minutes, until they are golden brown. Cool.

Max's grandma's french fries

Max, one of the French chefs with experience in Michelin starred restaurants in France and Switzerland, stayed with us for some time and cooked many fine meals. This recipe is his favourite family recipe. His Grandma used to cook it for him and I'll bet that's how he got inspired to be a chef. Something quite remarkable happens when you add just a smidge of curry to things – you don't get heat, just a mild spiciness.

SERVES 6
Time 30 minutes

1kg potatoes
2 cloves garlic
6 sprigs thyme
2 bay leaves
2 onions
100g bacon
1 tablespoon curry powder (Keens is fine)
Olive oil
Butter
Salt and pepper

1. Wash and scrub potatoes well. Slice in half lengthways then slice into wedges then dry on paper towel.
2. Generously coat the bottom of a large fry pan with the oil and heat until almost smoking. Add the potato wedges and turn them with tongs to coat with oil.
3. Add the garlic and herbs and turn the heat down a little. Turn the potatoes frequently until cooked.
4. In the meantime, peel and chop the onions and bacon into small dice. Heat the butter in a small pot and add the onions and bacon and fry gently.
5. Add the curry powder, salt and pepper and cook until the onions are translucent. Tip the onion mixture into the potatoes and stir gently to combine.
6. Cook for another 5 minutes or until potatoes are done to your liking.

Great with yoghurt or sour cream.

Pancake cake

If we had a dollar for every time a French house guest made crepes as their gastronomic contribution to our existence here on the farm... you know the rest! They seriously do love their crepes. Every single one of them, it seems, knows how to cook a mean crepe. These are not the thick pancake style we are most familiar with in the land down under. Rather, they are delicate and tender, and are just as delicious adorned with a drizzle of lemon and a sprinkle of sugar, or wrapped around some savoury morsels, such as asparagus in béarnaise sauce. A fun twist on the typical style of presentation was shared with us by Rudy and Roma, whereby the crepes are layered in a cake tin, with lemon curd smeared between them. Once the tin is full and chilled, the 'cake' is then cut into slices and served with a good dollop of thick cream.

SERVES 6
Time 1 hour

300g flour
2 eggs
¼ teaspoon vanilla essence
50g butter, melted and cooled
Approximately 250mL milk
1 large jar of lemon butter

1. Mix the flour, eggs, vanilla and butter in a bowl until very smooth. Gradually blend in some milk. Continue adding and blending until you have a runny batter. Cover and put the batter in the fridge for 3 hours or more, preferably overnight.
2. Remove batter from fridge and allow it to come to room temperature. Mix well before using.
3. Take a flat frying pan or crepe pan (if you have one) and put it on moderately high heat. Add about 1 teaspoon of butter and allow it to melt. Pour a small amount, about ¼ cup, of batter into the hot pan, swirling it around so it coats the pan with a thin layer. Allow to cook until tiny bubbles appear and burst on the surface of the batter, then flip the crepe over to cook on the other side. Remove and place on a plate covered with foil while you repeat the process with the remaining batter.
4. Line a spring-form pan with non-stick baking paper and place one of the crepes into the pan, cutting the edges to fit if necessary. Smear on 1 or 2 teaspoons of lemon butter. Repeat with the remaining crepes. Cover the pan with cling wrap then place a plate on top with a jar of jam or similar to weigh it down a little. Allow the pan to rest in the fridge for at least one hour. Remove jar, plate and wrap, then carefully remove the spring-form pan.

Garnish with icing sugar and serve with cream.

Avocado coffee smoothie

This would probably be put into the category of 'weird stuff that Murray eats' by our friend known as Mr T. Sure, it does sound a little different, but it is a really tasty smoothie and a good way to use up excess avocados. Poor us – what a problem to have, eh? With three large trees we are never short of these versatile fruit for more than a few months each year. We use them in place of butter, as an accompaniment to eggs for breakfast and in smoothies like this one – strictly for adults only.

SERVES 2
Time 5 minutes

1 frozen banana (or 1 fresh, plus 6 ice cubes)
1 double shot of espresso coffee
1 large avocado
1 tablespoon honey or less according to taste
100mL plain yoghurt
200mL milk

1. In a blender process the banana, coffee, avocado and honey until a smooth paste is formed.
2. Add the yoghurt and milk and process again until well combined.

Autumn

Is the time for... the annual harvest

The reward of the great effort that goes into tending the vines throughout the year is, without a doubt, the sight of giant tubs full of fully ripe, voluminous grapes. A climactic event, the annual harvest is both stressful and exhilarating. Will the good weather last? Will we get enough people to pick the crop? Does the winery have enough space? These and a thousand more questions flicker frantically through my mind at this time of year. Do I have enough food to feed to pickers? Will they pick all the grapes today? Is the quality good?

During the vintage period in many parts of Germany (and parts of adjoining countries, such as Austria and the Alsace region of eastern France), harvest festivals – to celebrate the pinnacle of the annual life-cycle of the vine and those whose livelihoods depend upon it – are a frequent occurrence. One of the main culinary markers of this time is the consumption of Zwiebelkuchen together with Federweisser – barely just fermenting wine (usually Riesling) that is a little bubbly, sweet, slightly alcoholic and almost milky due to the suspended yeast (hence the term that translates to feather-white or white as a feather). In Austria, it is known as Sturm and in Switzerland it is called Sauser.

Zwiebelkuchen (German for onion-cake) is a pizza-like creation typically made with fried onion, sour-cream and speck (or bacon). Two delightful girls from Stuttgart, Lene and Ines, introduced us to this culinary tradition. As a wine-maker I was astonished that I had never heard about the consumption of 'early wine', and checking with a number of wine-making colleagues in Australia, I realised I was not alone. Whilst we have adopted many customary wine-making practices, such as using different shaped bottles for different wine varieties to mirror those used in old-world winemaking countries, we have not captured many of the gastronomic traditions associated with wine that are still practised today.

Subsequent to our introduction to German gastronomy, Steffi and Olaf a German couple with whom we have developed a strong friendship, introduced us to the post-harvest dessert called Rote Gruetze. This concoction – in the Eastern parts of Germany – of red grapes and Sago is so popular that it is available through supermarkets, pre-manufactured. In other parts of Germany, the red fruit may be strawberries, black-currants, red-currants, or black cherries and the starchy component could be semolina or potato starch. Although it was a dish that I had never heard of before, more recently it became apparent that people in the Barossa Valley, in South Australia, were familiar with it due to the large number of German immigrants to that area many decades ago . None-the-less, it is unheard of amongst the general population of Australia. Since the introduction of these culinary gems, we have incorporated them into our lives, in part to celebrate the season, but also in memory of these lovely WWOOF guests.

It's lovely weather – warm, mild.
The air smells of faint, far-off tangerines with just a touch of nutmeg.
On my table there are cornflowers and jonquils, with rosemary sprigs...
How wonderful the Earth is.

JUANA INES DE LA CRUZ, EPISTOLA A FILOTEAMURRAY, JM. 1928. THE LETTERS OF KATHERINE MANSFIELD, VOLUME ONE. CONSTABLE & CO. LTD. UNIVERSITY OF CALIFORNIA.

Highlights of the season

Vineyard

In the early days of autumn much of our time is spent watching and waiting. Keeping an eye on the nets protecting the vines from those darn birds is vital to ensure we don't inadvertently end up with a very nice aviary for the little blighters, as they can easily squeeze their way in or under the net given an extra centimetre or two! We're also watching the grapes complete their ripening and once they start to sweeten up, I regularly test the sugar levels and the flavours to determine how soon we might need to pick them. I may have also conducted a yield estimate, which entails counting the number of bunches on a random number of vines, determining the average bunch weight then multiplying it out to give us an idea of how much fruit we have for winemaking. This gives Rob Diletti (crowned Winemaker of the Year 2014 by James Halliday) at Castle Rock Winery, some data for his vintage planning. He needs to know how many tanks to allocate to our wines, mostly the white varieties Riesling and Sauvignon Blanc. Sometimes we also send our red grapes to him for processing as we can only handle small amounts here on the farm with our miniature-scale crusher-destemmer and press. If we don't have much Shiraz or Merlot we don our winemakers' hats and do it ourselves, and we always crush the Pinot Noir here as it barely produces more than two barrels worth each year. Many woofers have been thrilled to be involved in the winemaking process, particularly at such a small scale. Prior to purchasing our crusher-destemmer, some even got to perform pigeage – foot stomping – a some-what sensuous experience, as the grapes are gently massaged by one's feet until they squish and slosh between one's toes. The house-rule is that one must have a glass of wine in hand whilst performing this feat. It's mostly fun!

The harvest is usually conducted in a bit of a flurry, and sometimes even a frenzy. It doesn't matter how much testing of grapes and predicting of ripeness we do, the vagaries of nature often throw a spanner in the works. A period of cool, cloudy weather slows down the ripening of the grapes considerably. A few hot days can hasten the process, and rain can just plain ruin it! Also, our grape picking requires people – no machine harvesting our precious grapes here, thank you! The logistics of getting enough people on a day when Castle Rock Winery can take the grapes, and when it is not raining nor blazingly hot, can sometimes be overwhelming. Then, there's feeding the troops! Part of the picking package we offer the groups in our community who are our usual pickers is that they receive hearty sustenance – as well as financial reward – in return for their efforts. A selection of home-made goodies, including some of Shirley's best cakes, pies and pasties, and fresh fruit are offered for morning 'smoko', whilst lunch is as varied as my imagination. Some years we've provided a hearty, hot meal; other years a barbecue; but usually it's a range of salads, cheese and freshly baked bread. And, more cake!

This time of year is also dotted with public holidays, often including Easter, so things can be quite hectic at the cellar door. Long weekends are prime time for city folk to embark on road trips to the south, so Perth experiences a mass exodus while Albany bulges at the seams with visitors. The annual Taste Great Southern festival is usually in full swing, and the Great Southern arm of the Perth International Arts Festival still has many musical, theatrical and creative events on show. The town is abuzz, and we enjoy many busy days serving our customers the fabulous selection of local produce on offer – with our delicious wine, of course!

Food

Figs, plums and grapes all reach their peak during the autumn. Some years, I feel as though I'm drowning in plums: everywhere I look there are plums – on the table, under the table, in the fridge, in the freezer. We make plum jam, plum paste, plum chutney and plum sauce. We give them away to friends, we sell them to our customers and we give them to the chooks to recycle into eggs! You get the plummy picture, don't you? The figs are a bit less prolific (thankfully) but are a favourite around here. They are particularly delicious on pizza, which is just perfect because Monday night is Pizza Night here on the farm and this is Murray's domain. Many of our woofers say that Murray's pizzas are the best they've ever had. Most of them have been involved in the process, either rolling out dough or preparing and arranging the ingredients under his guidance, so there's a certain sense of pride they enjoy too. It is quite a communal, bonding experience making pizza, and we're sure it's part of the magic that our woofers experience here at Oranje Tractor. And this is just from our domestic oven or our 'el-cheapo' gas-fired outdoor pizza oven. I can't wait until we build our own wood-fired pizza oven – another job for Glenn, the super friend and handyman *par excellence*.

woofers

Jonathan was lovely young French man who we were lucky to have woofing with us for a considerable amount of time. Unfortunately his visit almost resulted in a major calamity – one of very few accidents or injuries to have occurred on the Oranje Tractor Farm. Fortunately, it all ended well and we are the proud owners of some gorgeous furniture he made us. But to fill you in on the details let me just point out that Jonathan was not unfamiliar with carpentry – it was his occupation in France. Thus, I had no qualms about leaving him to work unsupervised on the task of using retired oak barrels to make some tables and chairs. I was in town, which is 10km away, when I got the call from Robby, our then vineyard hand, to say that Jonathan had drilled through his thumb. I heard 'brain' initially, in the crackle of the Blue-tooth connection in the car, so my heart was in my mouth to start with. Robby sounded a bit shell-shocked too so, not knowing how bad it really was, I asked him to drive Jonathan to the hospital, where I would meet him. I fully expected a gruesome sight, but was relieved to find that the drill had pierced only the fleshy part of his thumb, and missed the bone... although only by a whisker. I was actually more shocked when Jonathan refused to have a tetanus shot, and asked if this was because he had a needle phobia. Far from it! He explained how he had done most of his own tattoos – without anaesthesia – and that his mother was a fierce anti-immunisation campaigner back in France. I was terrified that he was going to get tetanus and die (drama queen hat on here) but the treating doctor reassured me that all would be OK. I don't know how he knew, but Jonathan did survive to live another day, and he's currently floating around some tropical island with Hitomi, the gorgeous Japanese girl he met while he was staying with us. Cute, huh?

Santee, Salute, Prost – it's all in the eyes

Hosting guests in one's home, particularly over a period of weeks, facilitates a degree of openness and frankness not otherwise possible between strangers. As a prime example, after two weeks with us, one of our French woofers delicately pointed out that not looking each other in the eye when clinking glasses and announcing Santee (or Salute, Prost or other equivalent of 'cheers' or 'to your health' in the relevant language) is considered poor etiquette. Further questioning – this time of German guests – revealed their belief that this lack of eye contact is likewise bad form, and dooms one to 'seven years of bad sex' or similar malady! Although this was considered humorous, it was a theme oft-repeated not only by French and German guests, but also, Italians and in fact, anyone from continental Europe. It became apparent that only the British and other new-world inhabitants, such as Americans and non-French Canadians, share Australians' poor form in this cultural ritual. It is perhaps no coincidence that these latter countries also share our propensity for binge-drinking. What has become very evident to us is that for Europeans generally, wine is consumed with the meal – not before, not after. In Australia pre-dinner drinks (particularly wine) is the norm for most people who drink alcohol in this country. A glass or two consumed whilst preparing the meal is a common habit, as is 'finishing the bottle'. Australia has an unenviable record of alcohol consumption and misuse. Perhaps it is no surprise given Australia's history, in particular the 'rum currency' of early settlement by a male-dominated (one to four ratio) convict society.

There is such a large and diverse repository of similar stories from our former guests – from the excellent coffee in Australia, to the woeful bread (according to the Germans and French); from the call for Aussies to linger longer at the meal table to enjoy some conviviality, to the unparalleled friendliness of the people of this wild land – that I mourn the lack of space and time to elaborate further. However, I trust that these examples have provided an insight into the wonderful exchange that being a WWOOF host has provided.

Autumn Recipes

121 Cinnamon rolls
123 Zwiebelkuchen
125 Apple butter
127 Pear, raspberry & chocolate crumble
128 Quince baked middle eastern style
129 Olive cake
131 Rote gruetze
133 Spicy pear chutney
135 Eldas fresh green olives
137 Japanese chicken curry
139 Caramelised sweet potato soup
141 Spaetzle
143 Pear and almond soufflé
145 Hannah's apple bake

Cinnamon rolls

There's nothing more Swedish than salmon and cinnamon rolls – but not together! Both were ubiquitous on my recent, first time visit to this truly beautiful country. I was lucky enough to visit in the Swedish summer, and, I was told, it was fortunate that the sun did shine because sometimes it just doesn't. In fact, I came home with a Swedish suntan – unintentionally, of course! My host, our wonderful ex-woofer Martin, was a fabulous tour guide and made sure I enjoyed some of the culinary treats that Sweden has to offer whilst we visited some spectacularly beautiful countryside and cityscapes. Eating lots of fresh salmon was a particular highlight for me, as was eating a real, Swedish cinnamon roll.

This recipe was actually cooked by a French woofer, Lauranne, but it is authentically Swedish, just like Ikea and Volvo! We were hosted by Lauranne and her boyfriend Guilliame at his family's mountain house near the spectacular Mt Blanc when we visited France a few years ago. On that occasion they treated us, among other things, to a fondue they cooked using the local cheese we bought earlier that day directly from the dairy. But don't let that happy memory distract you – these cinnamon rolls are fantastic too.

SERVES 8
Time 2 hours

Dough
150mL hot milk
50g butter
300g plain flour
6g baker's yeast
½ teaspoon salt
50g brown sugar
10 cardamom pods, crushed and husks removed, or 2 teaspoons ground cardomom

Filling
100g butter, at room temperature
50g brown sugar
1 tablespoon cinnamon

Glaze
1 egg
1 tablespoon water
3 tablespoons of pearl sugar (if you can get it!) or regular white sugar

1. Make the dough by mixing the butter with the hot milk until it has melted then allow it to cool to blood temperature. In a bowl mix the flour, yeast, salt, sugar and cardamom. Pour the milk-butter mix into the dry mix and form into a firm dough. Knead the dough for 10 minutes then shape it into a ball. Put the ball into a clean, oiled bowl and cover with a clean, damp tea-towel until double in size (about 1 hour).

2. Make the filling by mixing the butter with sugar and cinnamon with a fork.

3. Remove the dough from the bowl and roll it out to form a 30 x 30cm square, about 5mm thick. Spread the filling on the surface then roll up the dough, like a Swiss roll. Cut pieces 1cm thick across the roll and place in paper cake moulds on an oven tray. Set oven to 220°C. Allow dough to rise again for 45 minutes. Beat the egg with the water then brush this over the dough. Sprinkle sugar over the top of the buns and bake for 6 minutes. Alternatively, if you are in a hurry, don't wait for the rolls to rise, just put the tray into a cold oven and set to 200°C, then cook for 10 to 15 minutes.

4. Remove from oven and place rolls on a rack to cool.

Zwiebelkuchen

Two very capable young German women joined us just prior to our 2006 vintage and brought not just their good humour and helpful assistance. Ines and Lene were their names and I learnt from them that, just like France and Italy, Germany too has many fine food traditions – a gastronomy all of its own. Granted, it is not as well recognised, but I think given the increasing interest across the western world of eating 'tail to toe', German cuisine may soon experience a renaissance or at least an increase in awareness by those with an interest in food and gastronomy. After all, Germany is a large and productive country and has a long and rich history (apart from the little hiccup in the middle of last century). Agriculture began there after being introduced from the Middle East, and continues to be an important aspect of German culture. Known for its beer and the proverbial Oktoberfest, Germany also produces delicious grain breads, cheeses and, of course, wine. The Riesling variety is its most important, but other wine varieties are grown and some very good wines are also produced from them.

Pork is one of the most traditional meats because, until about 40 years ago, most families living outside of the big cities could raise a pig or two around the house, butcher it and store almost a year's worth of sausage, speck (which is like bacon) and other small goods made from the animals. Today, traditional food served in a gastrohaus usually features black pudding, brawn, pigs' trotters, rillettes and similar foods. Offal and bone-in meat cuts are a rich and concentrated source of nutrients and health nurturing compounds such as gelatine and collagen. These products are becoming popular among paleo-primal diet followers today and also among chefs and gastronomes.

This traditional harvest tart is like a thick, hearty pizza and has a lovely balance of salt (from the speck), sweet (from the caramelised onions) and sour (from the sour cream). The richness can be offset by drinking dry Riesling, or, as is traditional near the wine growing regions at harvest time, Federweisser.

SERVES 8
Time 60 minutes

Dough
400g flour
5-8g dried yeast
1 teaspoon sugar
½ teaspoon salt
4 tablespoons oil
250mL water

1.5kg onions
2 tablespoons butter
Salt & Pepper
1 teaspoon rosemary
1 teaspoon cumin
1 clove garlic

Topping
350g streaky bacon (or speck, if you can get it)
200g Gouda cheese
3 eggs
2 tablespoons sour cream or crème fraîche

1. Mix the dry dough ingredients in a bowl. Add the oil and water and mix until combined. Knead for 5 minutes then allow to rest in a warm place until risen (about half an hour).

2. Peel, quarter and slice the onions. In a large fry pan, melt the butter then add the onions with salt and pepper, rosemary and cumin. While this is gently frying, peel and mince the garlic clove then add it to the frying pan. Simmer the onion mixture for 15 minutes or until all the liquid has evaporated. Turn off heat and allow the mixture to cool. Grease a large baking tray.

3. Grate the cheese and dice the bacon. Add these to the cooled onions. Roll out dough to fit the baking tray. Put the dough onto the tray then top with the onion and bacon mixture. Allow to rise again for 20 minutes then put into oven. Turn oven on to 180°C. Bake for 40 minutes.

Apple butter

For much of the year, apples adorn our tables and feature in our food choices. Murray planted 40 different, mostly heirloom, varieties of apples originally. There's been some natural attrition, mostly due to drought conditions a few years ago, but we still have tree-ripe, fresh apples from mid-December until July each year. If you know anything about growing apples, you'll know that the usual season is a lot shorter than this. Of course, apples store particularly well in cold storage, so supermarkets sell apples all year. But fresh, really fresh, apples are usually autumn ripening. Murray was able to source some summer ripening apples to extend the apple season. Why? Partly because he loves nothing more than to chomp on a crisp, crunchy apple, and partly because of his family's history as apple orchardists in Mount Barker, not far north of Albany. And partly because he could!

We'd never heard of this great way of preserving the glut of apples we get annually, until Max, one of our industrious German woofers showed us.

Makes 4 medium jars
Time 2 ½ hours

6 apples, peeled and quartered
180mL apple cider or juice (unsweetened)
2 to 4 tablespoons sweetener of choice (sugar, honey, stevia, etc)
1 teaspoon cinnamon
¼ teaspoon ground cloves

1. Peel and core the apples and cut into chunks.
2. Add the cider or juice to the apples in a pot together with spices and sweetener.
3. Bring to boil, then reduce heat, cover and simmer for 1 hour, stirring often.
4. Remove cover and simmer until thick and dark. Continue to stir frequently.
5. Bottle into sterile jars and seal.

Pear, raspberry & chocolate crumble

Flo and Remy were a gorgeous French couple who not only spent some time with us, but sent Flo's parents the following year. It was easy to see where she got her lovely nature from. We thoroughly enjoyed her parents' company despite our virtually non-existent French and their admirable but limited English.

Flo and Remy introduced us to a concept we hadn't experienced before: that of individualised serviette rings. Apparently most family homes in France have a set of such rings so that one doesn't have to launder the serviettes (the linen variety, not the paper ones) after every family meal, only when they are no longer clean enough. That doesn't apply to guest serviettes, of course, but it does make good sense for a family, especially if you want to reduce your carbon footprint. We have adopted this strategy at our dining table because most woofers expect a serviette with their meal.

This yummy crumble, whipped up by Flo, can be made without the chocolate, or with dark chocolate instead of white.

SERVES 6
Time 45 minutes

4 ripe pears
250g raspberries
60g butter
50g flour
40g ground almonds
50g brown sugar
50g chocolate chips, white or dark

Set oven to 180°C. Butter an 22cm ovenproof dish.

1. Peel and core the pears and chop into 2cm chunks, then place into the ovenproof dish.
2. Scatter the raspberries over the pears.
3. Make the crumble by putting the flour, almonds and sugar into a bowl and then adding the melted butter.
4. Mix well then add the chocolate chips and stir gently.
5. Spoon the crumble over the fruit and bake for 30 minutes or until the crumble is starting to brown slightly.

Quince baked middle eastern style

In the early years of our WWOOF life, we hosted a number of Israelis. One young woman, who had spent much of her life in the USA, was well informed about the culture and history of Jewish food and this was fascinating. We were also quite intrigued that she ate vegetables for breakfast – salad style – and said that this was not uncommon for Israelis. It may seem odd at first, but it is a healthy way to start the day. Much of the food of Israel has Middle Eastern influences, such as this gorgeous dessert based on quinces and dates.

2 quinces
200g pitted dates
100g unsalted pistachio nuts
100g roasted almonds
1 tablespoon honey
Grated zest of 1 lemon
1 tablespoon orange blossom water
¼ teaspoon cinnamon

1. Set oven to 160°C.
2. Rinse the quinces and rub the fur off the skin.
3. Cut in half cross-wise, remove the core to leave a small hollow and place, cut side up, in a deep casserole dish.
4. Chop the dates roughly and put them in a small bowl with the nuts and zest. Mix well, then add the honey, orange blossom water and cinnamon, stirring to combine.
5. Put the mixture in the hollows of the quinces. Put about 1 cup of water around the quinces and cover the dish with foil.
6. Bake for 45 minutes or until the quinces soften.
7. Remove the foil, turn oven up to 180°C and bake for about 10 minutes or until the nuts start to colour.

olive cake

This is a very delicious, quite historic cake. Magali, a young woman from the south of France, introduced us to the joy of this moist loaf and to its background. A typically Provençal apéritif, Cake aux Olives, is often kept on hand 'in case someone drops by'. It might be friends; it might be the President – regardless, it is usually cut into cubes and eaten with the fingers, and always served with wine. It can be served with a salad for a light lunch, too – knife and fork optional! Olive cake should be made in a loaf tin, rather than a round one, for tradition as much as cooking efficiency. In the French language 'un cake' describes the square or oblong shape of the tin. Cakes cooked in a round tin are called 'un gâteau'. Savoury cakes like these are very easy, quick and adaptable – although I think one always needs olives. Try adding smoked chicken or fish (Bouverie Smoked Trout, a great product from Mount Barker, for example) or even some chopped, roasted vegetables and thyme. You can also include grated tasty cheese, such as Gruyère, and omit the nuts.

SERVES 8
Time 1 hour

250g flour
3 teaspoons baking powder
Pinch each of salt and pepper
200g olives, seeded and chopped
150g ham, sliced into matchstick-like pieces
50g hazelnuts, chopped roughly
4 eggs
100mL white wine
250mL olive oil
100mL port

1. Set oven to 180°C.
2. Grease a 20cm loaf tin.
3. Sift the flour into a bowl and mix in the baking powder, salt and pepper.
4. In a separate bowl, stir together the olives, ham and nuts then add the eggs, wine, oil and port.
5. Add the wet mix to the dry mix and stir well to combine, then pour into the loaf tin.
6. Bake for 45 minutes or until cooked. Check using a skewer – if it comes out clean, it's done.
7. Allow to cool for 10 minutes in the tin, then turn out onto a rack to finish the cooling process.

Rote gruetze

Like many family-style recipes, this one is incredibly easy and amazingly delicious. Of course it has to be made with seasonal, fresh fruit, in this case grapes, but any red fruit can be substituted. We now make this every year during the grape harvest in autumn since being introduced to it by our dear German friends Steffi and Olaf in 2009. They hail from the very eastern side of Germany, right on the Polish border, and were woofing with us for several months before making their way around Oz in a beat-up van. Olaf took to Australia like a duck to water, and was able to talk Steffi into applying for immigration. We're still waiting for them to arrive. I think the two 'tin lids' they created whilst waiting for their visa might have thwarted that plan a bit!

It almost goes without saying that the tastier the fruit, the better the finished product. If you can possibly get ripe wine grapes, choose those over table grapes because they have so much more flavour and colour. You must have red grapes because rote is German for red, after all!

SERVES 4 to 6
Time 40 minutes

1kg red (black) grapes
100g Sago

1. Crush the grapes either with your hands over a jug or in the Thermomix on reverse speed 4 for 30 seconds.
2. Strain the crushed grapes into a pot, pressing as much juice out as possible. You should have about 700mL.
3. Add the Sago and heat the pot gently, stirring frequently until the Sago turns translucent. This takes about 20 or 30 minutes.
4. Remove from stove and allow to cool a little.
5. Pour the finished mix into serving bowls (glass is preferable) and allow to set in the fridge.

Serve with cream.

Spicy pear chutney

When we have filled the freezer and stocked the shelves full of poached or preserved pears, this chutney enables us to put the extras to good use. We enjoy chutney with our cheeses, and seasonal chutney is always present on our Tractor Driver's Platters in the cellar door. Oddly it is only we folk from the former British colonies that eat and enjoy chutneys these days. The French are, in general, a little perplexed – why would you want to mask your cheese in strongly flavoured, savoury preserves? Good question, and understandable when you have hundreds of beautiful cheeses to choose from in France. But when you've only got cheddar... !

The Germans too find our penchant for chutney somewhat intriguing. It probably stems from the fact that chutney made its way to the Australian table via the British Empire, which had in turn adopted the Indian custom of creating chutneys for serving with curries. It surprises everyone to hear that it is thought that the concept of chutney probably originated in Northern Europe in about 500 BC as a way of preserving fruit and vegetables. Napoleon was said to have kept his army going on good stores of chutney, and Captain Cook's ability to keep his crew free of the dreaded scurvy is likely due to the use of generous amounts of lime pickle (the lime being rich in vitamin C, and the reason the English were often called Limeys in the USA, rather than Poms as by us Australians). For some reason, chutney is no longer popular in continental Europe.

What to do with chutney? Apart from enjoying these sweet-sour-spicy pastes with cheese, we use chutneys as a great accompaniment to cold meats, in sandwiches with cheese or meat, and even for marinating raw meat prior to cooking, or instead of tomato sauce on a pizza. One of our English woofers, Paul, made us a big batch of this fiery chutney. It is now a favourite.

Makes 6 medium jars
Time 1½ hours

1 red onion
2 garlic cloves
4 small chillies
1 tablespoon olive oil
½ teaspoon ground cinnamon

2 star anise pods
Pinch of salt
300g sugar
1kg pears
125mL apple cider vinegar
Juice of 1 small lemon

1. Peel and core the pears and chop into 1cm cubes. Peel and halve the onion and garlic. Remove the stem and seeds from the chillies.
2. Put onion, garlic and chillies into a blender and blitz until finely chopped.
3. Heat oil in a pot and gently fry the chilli mixture for 5 minutes or so. Add cinnamon, star anise and salt.
4. Add chopped pears to the pot along with the sugar, vinegar and lemon juice and stir until sugar dissolves.
5. Allow the mixture to boil gently until thick, about 45 minutes.
6. Ladle the chutney into hot, sterilised jars and seal with a lid.

Leave at least one month before opening to enable the flavours to mellow and blend.

Elda's fresh green olives

One quiet, early autumn day a few years back, a small car crept gingerly up our snaking gravel driveway. Ready to proclaim that we were actually not open this day, I approached the woman as she uncrumpled herself from the vehicle. She was an older woman – which threw me a little, as most of our typical customers are in a younger age bracket – and before I could get in with my rehearsed spiel, she started pointing at the olive trees lining the driveway. She introduced herself as Elda and explained, in her delightful Italianesque English, that she wanted to buy some of our plump green olives. I was thinking she might want the whole crop, so was a little hesitant at first. However, she quickly made it clear that she only wanted a bucket of them. Instead of asking for payment, I suggested she share her method for dealing with the olives with me, as I had almost given up trying to do anything with green olives, preferring to wait until they blackened so that I could salt-dry them. Elda was more than happy with this arrangement, and not only did she explain precisely her process, but she also dropped by three or four days later with a delicious batch of these freshly cured olives. They don't last long, probably at most two weeks in the fridge, but if you can score yourself some green table olives, it's well worth having a go at this. Your taste buds won't regret it!

1. Smash each olive between two pieces of wood, or using a beer bottle on a wooden chopping board.
2. Remove the seed and place the olives in a large kitchen bowl.
3. Pour enough boiling water over to cover and add handful of salt.
4. Put a lid over the bowl and leave on the sink.
5. The next day, drain and wash the olives then refill the bowl with cold water and a slightly smaller amount of salt.
6. Every day for 4 days, repeat this procedure. Taste the olives to determine if they have lost enough bitterness for you. If so, drain and dry the olives on paper towels.
7. Put the olives into a large glass jar. For every 1 litre, add the juice of 2 lemons, one teaspoon of salt, ½ cup olive oil, 2 cloves of crushed garlic, 2 chillies, the leaves off one sprig of rosemary and one of oregano, 1 bay leaf and 6 peppercorns.
8. Refrigerate and use within the following month, if they are not eaten sooner!

Warning: they are so delicious, you will not be able to stop at one!

Japanese chicken curry

In the early days of being a WWOOF host, we enjoyed the company – and the cooking – of many young Japanese women. We both love the freshness of Japanese cuisine, so despite the frequent language difficulties (my high school Japanese having long since vanished from my memory banks) we enjoyed hosting these gutsy gals. Mostly, they would have participated in a three-month English language course in Perth before heading off – usually on their own – to put their newly learned skills into practice. Woofing was an obvious choice, since it was virtually free and a cultural adventure at the same time. As anyone who's spent time with Japanese people will know, they are some of the politest people on earth. Consequently, they have a hard time saying 'no', which resulted in many a Japanese girl telling us they had gained 10kg of weight or more during their short time in Australia. Although we thoroughly enjoyed almost all the meals cooked by our Japanese guests, we were astonished that even the most accomplished among them used a box of Japanese Curry Sauce to create Chicken Curry! All except Kayoko. Here's her version, and a copy of the instructions, complete with pictures!

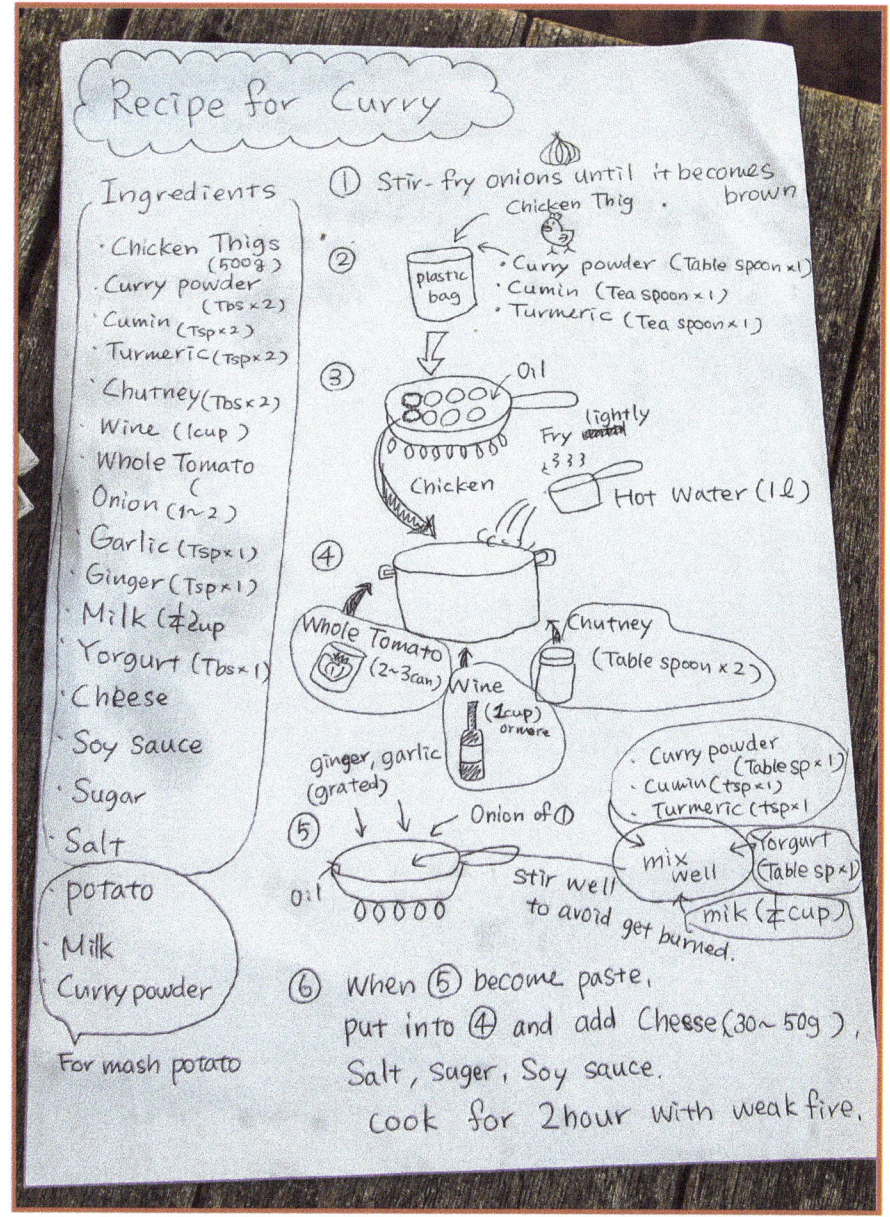

Caramelised sweet potato soup

This soup has become an all-time favourite for both us and our woofers. Having to constantly feed people who are working for you means that you need quick, easy and tasty recipes on hand. This recipe is a shortcut from my favourite geeky cooking resource called 'Modernist Cuisine At Home' (www.modernistcuisine.com) which combines state of the art techniques with relevant, current science to help you to create amazing food. The original recipe is fine for a dinner party, where extra effort may be deemed appropriate. Mine is deceptively easy, but you will need a pressure cooker. The result is a decadently rich tasting soup that belies the simplicity of its ingredients and method. You can interchange parsnip, carrots, corn or even red capsicum with the sweet potatoes and you will still get a delicious caramel flavour.

SERVES 4 to 6
Time 30 minutes

2 to 3 large sweet potatoes
100g butter
1 teaspoon salt
¾ teaspoon bicarbonate of soda
1L water

1. Peel the sweet potatoes and chop into roughly 3cm pieces.
2. Place in the pressure cooker and add the rest of the ingredients. It is imperative that you use ALL ingredients in the amounts stated.
3. Put the lid on the pressure cooker to seal, set to high pressure then turn on the heat. Once the cooker reaches pressure, start the timer.
4. Once 20 minutes is up, turn off the heat, remove cooker to the sink and allow to cool until the pressure and safety valves indicate it is safe to open. Alternatively you can run cool tap water over the cooker to cool it down more quickly (if you're in a hurry).
5. Transfer the contents of the cooker to a blender – probably best in two batches – and blend until very smooth, adding water to thin until your preferred thickness is reached.
6. Reheat, if necessary, and serve with a dollop of yoghurt or sour cream and some crusty bread.

If you're interested in the science behind this, you can download a free sample of Modernist Cuisine at Home (which includes the recipe, from which this is derived, for caramelised carrot soup) which explains it far more eloquently than I could.

Spaetzle

I had no idea Germans, particularly the Bavarians and the Swabians, ate a lot of noodles. Italians, yes; Chinese, yes; but Germans? No, yet they have also been eating them for centuries, according to Sabina, one of the first German woofers we hosted. Invented in the Swabian region of southern Germany, where Sabina was from, Späetzle are rustic egg noodles which are also soft and light. They are quick to make, but require a bit of practice to get the technique right – although it really doesn't alter the taste if your Späetzle are rougher than rustic. They are not meant to resemble the smooth, neat noodles of Italy or China. When I visited Sabina and her parents during my Churchill study tour of European organic wineries, her mum cooked Späetzle for me too – and it was just as good as Sabina's, of course. She served the noodles layered with grated cheese and roasted onions and called it Käsespäetzle. Alternatively serve the noodles alongside roasted meat, or go non-traditional and add your favourite Italian pasta sauce.

SERVES 4 to 6
Time 30 minutes

400g flour
1 teaspoon salt
4 eggs
3 tablespoons water plus
About 125 to 150mL water
About 50g butter

1. In a large bowl mix the flour with the salt.
2. In a small, separate bowl beat the eggs with the small volume of water. Pour the eggs into the flour and stir to combine thoroughly, then add the larger volume of water and mix for 5 minutes. You will have a runny dough, almost like a batter.
3. Bring a big pot of water to the boil and add a teaspoon of salt.
4. Moisten a small cutting board with hot water, then pour a few tablespoons of the dough onto the wet surface.
5. Hold the tray over the pot of water and start slicing the dough (using a long blunt knife or a spatula) into thin strips, dropping it into the water below. Or you could get yourself a special tool, as my friend and editor, Penny has done.
6. Allow this to cook until the noodles rise to the top of the water.
7. Scoop out the cooked noodles and rinse them under cold water in a colander or sieve.
8. Set aside and repeat with remaining batter until it is all used. Heat the butter in a large frying pan then add the noodles, warming them and causing them to puff a little.

Serve immediately.

Pear and almond soufflé

Recently, we've had a tsunami of French chefs wanting to woof with us. Not just any old French chefs, but ones who have worked in Michelin-starred restaurants in Europe. To say this has been fantastic is truly an understatement. These guys (and they are all male) share an interest in fresh, seasonal produce and they are passionate about cooking. Many a fine meal has been whipped up under our roof of late. This soufflé is one such example, and fortunately it is dead easy, unlike a lot of the other meals we have enjoyed. I like this one because you can interchange with other poached fruit and nut combinations. Quince and macadamia is a great alternative.

SERVES 4
Time 45 minutes

2 pears
50g sugar
250mL water
60g coarsely ground almonds
30g butter, softened
1 tablespoon white sugar
4 egg whites (not a speck of yolk)
Very small pinch of salt
40g white sugar

1. Peel and core the pears. Make a syrup from the sugar and water by heating gently in a pot until the sugar has dissolved. Place pears into the pot and simmer about 15 minutes or until the pears have softened.
2. Allow to cool, remove from syrup, then purée the pears with 60g coarsely ground almonds. Weigh out 160g of the purée.
3. Using a pastry brush, paint the inside of 4 ramekin dishes with the butter, using upward strokes starting in the middle of the base. Sprinkle about 1 teaspoon of sugar in each ramekin and rotate it to get the sugar coating the inside walls. Gently shake out any excess. Place in refrigerator until required.
4. Using a very clean and dry bowl, whisk the egg whites with salt until they are firm. This could take as long as 20 minutes by hand, so don't give up. Start adding the sugar a teaspoonful at a time, whisking well between each addition, until all sugar has been incorporated.
5. Now, fold in (mix in very gently) the pear and almond purée until just combined. It's better to have a bit of white visible than to over-mix it.
6. Remove the ramekins from the fridge and pour the mix into each, stopping a centimetre below the rim. Using your thumb and forefinger, pinch the top of the rim and run your hand around the top, removing the butter and sugar in an even manner. Don't skip this, it is critical to get the soufflé to rise straight.
7. Put the ramekins into the fridge for 20 minutes. Set the oven to 180°C. Bake the ramekins for 10 to 15 minutes.

Eat immediately.

Hannah's apple bake

Hannah was a beautiful young woman from Germany who captured our hearts with her charming and intelligent disposition. She cooked this simple entrée for us and its simplicity belies its enjoyment factor. There's a range of flavours and textures here that will make you want more.

SERVES 2
Time 20 minutes

2 apples
1 fresh goats cheese, eg. Ringwould Frais – approximately 100g
1 teaspoon honey
2 sprigs thyme
1 sprig rosemary

1. Set oven to 180°C
2. Slice the apples across their middles into 4 or 5 pieces and discard the top and bottom bits, leaving uniform round apple disks.
3. Slice the cheese into rounds. Starting with a slice of apple on the bottom, layer cheese slices between the apple slices, finishing with a cheese slice.
4. Place the apples in a small ovenproof container.
5. Drizzle honey over the top and sprinkle with the herb leaves.
6. Bake for 15 minutes, or more, until the apple has softened and the cheese is golden on top.

This could be served as an entrée, an accompaniment to roast pork or as a dessert.

To make a good salad is to be a brilliant diplomat
— the problem is entirely the same in both cases.
To know exactly how much oil one must put with one's vinegar.

Winter

Remember, revive & rest

Once the grapes have been harvested, and the autumn fruit from our orchard picked and pickled, or preserved, there's not a lot to do – except wait for the leaves to drop and senescence to settle in. It is the time of rest for the vines, for most of the fruit trees, and even for us. This is the time that we have time to take a break, recharge our batteries and prepare for the year to come. Sometimes, our break does not occur until after we have pruned the vines, around September. This is not such a bad thing because we can often head north – in Australia, or way north – to the other hemisphere – to chase some sun and some adventure in places different from our own. Travel is, and always has been, a great personal reward and a great learning opportunity for me. Sometimes it is merely about learning about myself; other times it is learning about different cultures, cuisines and climates. From Austria and Germany, through to Bali and Cambodia, the USA and Canada; whether it is cycling, walking or driving; eating, drinking or sight-seeing; I've gained more than I could possibly explain by travelling. Travel is the great educator, the great humbler, and the greatest irony considering we live in what many may see as paradise: a hand-built climate-sensible, passive solar house; a beautiful organic vineyard; a flourishing orchard and vegetable patch; the potager garden and the rustic cellar door and café, all within 10 minutes of a regional town, with all its amenities, on the pristine south coast of Western Australia, far from the madding crowd, far from toxic industry, far from traffic lights! Yes, the pride of Albany – no traffic lights. It must be one of the only towns (or cities) of its size to be able to lay claim to this rather amazing fact.

One of unexpected benefits of hosting guests under the WWOOF scheme is the generous and warm welcome we have received from our former visitors in their own towns and cities, and often in their own homes when we have visited these countries. It has been a privilege to see places from the perspective of a local resident rather than through a tourist's eyes. Our experience, particularly in Europe which we have visited a few times over the past decade, has been a bit like couch surfing – only better. And it's a bit like visiting family – only better! There's a strong bond and a common history between us, which is not marred by the baggage that sometimes occurs with families. We have also seen many sights and enjoyed many experiences that the average Aussie tourist probably wouldn't. For example, the Mercedes Benz museum in Stuttgart, a wonderful example of German ingenuity; a boat trip on the Oder river in the former East Germany which included an enormous 'boat lift' that moves a giant swimming pool-like structure loaded with barges and boats up and down the equivalent of four or five storeys, whilst we were still in the boat; a wander through some of the Traboules – 4th century covered, and often hidden, corridors in old Lyon, where precious silk was transported between the loom-houses and the river boats for transport to distant places; a mountain walk in the French Alps in early autumn, nibbling on wild berries on the way and culminating in a pique-nique overlooking the magnificence of Mont Blanc as the sun shone on the high peaks; a visit to the Night Restaurant in Berlin, where many of your senses are challenged as you eat your meal in complete blackness, served by blind wait-staff... I could write a small book about the fabulous fun we have had during these times, but as Murray reminded me 'one book at a time is enough'.

Highlights of the season

Vineyard

For all intents and purposes the trunks look dead, being devoid of anything green, during the winter months of July through to September. The past year's leaves have fallen off the vines, having performed their duty of creating energy from the sun which initially is used to grow and ripen the fruit, then to generate energy for storage in the roots of the vine for the following year's bud-burst. They form a crunchy, multi-coloured patchwork on the ground beneath our feet as we pass through on our daily walks. Merlot the dog is often sniffing underneath the piles of leaves for evidence of rabbits or rodents as he frolics along the rows. Until the canes are pruned, they form a framework upon which many a spider spins her web, glistening with early morning dew. Before long it is time to start the mammoth task of pruning. Each of the many thousands of vines must have its previous year's growth – often metres long – cut back to a short stump containing two spurs (buds which will, hopefully, burst in the spring and form the next year's shoots and fruit). In addition, any dead wood and surplus shoots are also removed. The prunings are collected and piled up out of the vineyard to be burnt, eventually. Our main purpose is to minimise the carry-over of disease – the mildews, mostly – and to enable us to have a bonfire or two later on. In the traditional wine growing regions of the world, the burning of the canes is a festive time and marks the changing of the season.

Woofers

The pictures paint this story, don't they? Thanks to Sabine, one of our most treasured woofers, and a few others, the pit in the garden where our rainwater tank once stood was transformed into a verdant labyrinth of herbs, salad greens, vegetables and flowers – otherwise known as a potager garden. Potage is French for soup and a potager kitchen garden is usually located close to the kitchen so the cook can zip out and grab a handful of fresh produce with which the soup is made. And, initially, it was a precisely planted potager! Sabine meticulously dug planting beds, creating a mosaic from the sandy soil, some rich compost and finally a layer of mulch around the individual herbs, flowers and bushes she and Helene (a French woofer) planted. Now, it looks more like a sprawling jungle most of the time. Shirley tries to keep it tamed with her periodic weeding, clipping and replanting, but nature is naturally wild and rampant, and nature often prevails.

Like most of our German woofers, Sabine's work ethic was particularly strong and most days we almost had to beg her to stop working beyond the four-hour WWOOF requirement. During the creation of the potager, however, there was no stopping her. She worked until sundown each day until it was done. And that was after completing myriad other tasks such as helping us set up the cellar door and kitchen. 'Can I dig now?' she would ask each day... and, boy, did Sabine dig with gusto. We shudder to think how many hundreds of kilograms of dirt she moved in creating the network of garden beds that needed to be filled with rich compost prior to planting. And, yes, she filled the beds and did the planting. Yay, Sabine!

Once that project was complete, Sabine embarked on another major undertaking – the creation of the iron flowers that comprise the sculptured wall in the garden. Often enveloped by a vigorous choko vine, this structure protects the vegetables from the strong south-easterly winds in summer and also serves to delight customers as they disembark from their cars in the vicinity. This creative endeavour required learning the new skill of welding, which she clearly mastered as you can see from the photos.

Sabine generously hosted Murray and me on our 2011 visit to Europe at her apartment near Stuttgart. We were on our way through Germany between cycle trips along the incredible cycle-ways of the Netherlands and then around Lake Constance, bordering south Germany, Switzerland and Austria. Sabine showed us around the architecturally significant city of Stuttgart, and also took us to Constance in her car via a 500 year old farm. This historic place has been refurbished using traditional methods, including thatching on the roof that was almost a metre thick.

Germany is vastly under-recognised as a tourist destination, in my humble opinion, as is German hospitality. When travelling, Germans can initially appear somewhat stoic, but I think that comes from having to share your patch with so many millions of others in close proximity. I can see why people just shut the rest of the crowd out and get on with their daily lives. However, our experience with the German guests we have hosted and the visits we have made to them and their families in Europe provides a radically different perspective. Warm, generous and articulate, they have all made very good company both as hosts and as guests. They are great travellers, and have a world view that is to be admired. On my most recent expedition to Europe, I visited the home of one of our youngest, but most delightful woofers, Alex, and felt immediately at ease such was the warm welcome he and his parents gave me. Actually, it was a bit embarrassing because they insisted on paying for absolutely everything – meals, local train fares, exhibit entry fees – as well as taking me to some gorgeous places and feeding me many delicious meals, accompanied by great local wines. I left hoping that they would visit us soon so that we could repay their generosity and thoughtfulness.

Food

I used to have my best thoughts in the shower. Aaaaah, those were the days, the BTF (Before The Farm) days when – being on the town water supply and not solely reliant on the rainfall collected from the roof of the farm buildings – meant one could luxuriate in long, hot showers with no concern about depleting one's drinking supply. Now, however, my thoughts come thick and fast on the twice daily tour of duty around the vineyard with Merlot the dog. I call it our hunting and gathering time. He, Merlot, does most of the hunting – for anything that moves, but especially those rascally rabbits – while I gather several things in addition to my thoughts. Sometimes I'm physically gathering grapes to taste and test, occasionally wild mushrooms; and other times I'm gathering evidence that all is good in the vineyard – or not. Frequently there's evidence of kangaroos and of Mr Reynard, the fox, who likes to eat our chooks for breakfast, lunch or dinner. And it's not just the evidence of what he or she ate (which is often grapes and beetles) that we find. Many meanderings reveal more than just a glimpse of the wily fox, and sometimes of more than one fox, face-to-face. One morning (not particularly early in the day) I let Merlot off the lead because he was particularly ecstatic about something. I thought it must have been rabbits given his shrieking performance, but was surprised to see, a few minutes later, two foxes being chased down the valley by Merlot. As I really didn't want to take him on yet another vet visit, I headed off after them to try to circumvent Round One of Mr and Mrs Fox versus Merlot the Dog. Very soon thereafter, Merlot was charging back towards me, and I noticed he had something in tow. As he got closer to me, I was astonished to see one of the foxes chasing Merlot. I figured that as soon as the fox saw me, it would high-tail it out of there... but it didn't seem to notice me until I started waving my arms about madly. Hell, I didn't want to be chased by a bloody fox! It quickly about-faced and headed back to the place from whence it had come – with Merlot in tow yet again. I imagined that this Jack Russell – Fox merry-go-round might go on all day, but thankfully Merlot gave up after a few minutes and returned to me, none the worse for wear, a little puffed but quite pleased with himself.

Murray's hunting and gathering is otherwise known as shopping in the garden. More often than not, we decide what to eat for dinner just before we cook it, so being able to pop outside to collect ingredients is very handy. Actually, I think that it's because we know we usually have a garden full of produce that we don't do a lot of planning around meals. This works just fine when there's just a few of us, but when there are three or four woofers or more to feed which sometimes happens, a little more organisation is desirable, and perhaps necessary. But I'm happy to say that no-one has ever gone hungry on the Oranje Tractor Farm. There's nearly always an abundance of fruit; in winter there's a fine selection of citrus fruit, from mandarins and oranges to lemonade lemons and grapefruit. As we bake our own bread, almost daily, and make our own yoghurt we can offer our guests a range of tasty treats to keep their stomachs from growling. And there's always loads of marmalade to pile on top. The only trouble is, for most Europeans, marmalade is the term given to jam of every description so they're often a little surprised that all the marmalade is citrus-based, and it seems that only the English and Scottish are particularly keen on it. Luckily, I've discovered that a jar of marmalade can be turned into a fabulous orange and almond cake in no time at all. While talking citrus, it has delighted us seeing the delight on our European guests' faces when they see lemons and oranges growing on the trees in the front yard. For many, it is the first time they've seen them on a tree. Same goes for avocados, but to an Aussie that's not nearly so surprising because most of us (well, of my vintage) grew up with a lemon tree in the back garden. Avocados, on the other hand are relatively novel, and backyards are now way too small in our modern Australian cities to grow one of these jungle giants! Many of our cellar door visitors, most of whom are Aussies, are likewise enthralled to see the large, plump green fruit hanging from the trees that line the pathway to the entrance. We start picking the avocados at this time of year, and continue for a good portion of the rest of the year enjoying them in abundance, even at breakfast with poached eggs or in smoothies. Murray's favourite smoothie is one we discovered in Bali a few years ago: coffee, banana and avocado. See page 110.

Winter is surprisingly good for growing lettuces too. And, since we've had so many French guests, we are no longer salad dodgers in winter. We are happy to enjoy fresh salad year round, just like they do… complete with a vinaigrette, often made quite simply with just our own olive oil and some of our red wine vinegar.

There are some fabulously creative people here in Albany, and we've been fortunate enough to enjoy the fruits of their labour, none more so than Rebecca Weadon from Croker Lacey Graphic Design. Rebecca and I met when she first arrived in Albany about a decade ago. We were participating in a small business workshop, and it transpired that she was pondering whether to set up her own graphic design business, having recently moved from Adelaide. Not only do we now belong to the same book club, but Rebecca has exhibited some of her artwork at the cellar door during the Great Southern Art Trail, and her design skill has been employed for a number of projects we've been involved with locally. In addition, the cellar door floor is adorned with her fabulous pastel creation, which reflects both the model of the tractor and the locavore aspect of our food offerings: 513 kilometres radius is where we source 95% of our food from. And you will, no doubt, note that the design work on this book is by Croker Lacey.

Creative community

Winter
Recipes

- 159 Provencal fig cake
- 161 Roti orlott
- 163 Steamed chicken Japanese style
- 165 Austrian apple struedel
- 167 Cumquat preserve
- 169 Apple tarte tartin
- 171 Anglesey eggs
- 173 Swiss choc mousse
- 175 German ravioli
- 177 Jewish apple cake
- 179 Zopf
- 181 Orange & almond cake
- 183 Sardines in vine leaves
- 184 Zito

Provencal fig cake

We have a number of fig trees and, although they are small, we usually have more fresh figs than we know what to do with in late summer. Eating fresh figs is something I assumed everyone had enjoyed, as I had throughout my childhood, and I was shocked when countless guests revealed they had never done so even once. There are only so many fresh figs one can eat with one's yoghurt and nuts for breakfast, as a mid-morning snack, on pizza or pan-fried with balsamic vinegar and served with blue cheese. Sure, you can make them into fig jam, but I think the best way of preserving them is to dry them in a very low oven or a food dehydrator, then store them in the freezer or preserve them in a spiced sugar syrup. Either way, you'll have figs for this fabulous cake that Jonathan the carpenter brought to us from the south of France. With the aroma of the rosemary and figs that wafts out when you cut through a slice, you'll be transported to Provence.

SERVES 6 TO 8
Time 1 hour

Topping
150g white sugar
90g brown sugar
80g butter
1 lemon, juiced
12 dried or preserved figs
1 teaspoon finely chopped rosemary leaves

Cake
250g butter
200g brown sugar
4 eggs
250mL cream
360g flour
2 teaspoons baking powder
2 teaspoons ground ginger
2 teaspoons ground cinnamon
½ teaspoon ground or grated nutmeg

1. Set oven to 180°C.
2. Line a 26cm round cake tin with non-stick baking paper.
3. Make the topping by melting the butter in a pan over medium heat then adding the sugars. Stir, then add the lemon juice, mix again and bring to the boil. Turn down heat to allow the mixture to simmer until it turns golden. This may take between 2 and 5 minutes.
4. In the meantime cut the figs in half lengthways. Once the caramel is golden, add the figs, cut side down and allow to cook another 3 or 4 minutes. Pour the entire contents into the cake tin and carefully (they'll be very hot) arrange the figs over the base.
5. To make the cake batter, cream the butter and sugar then add the eggs one at a time, beating between each addition. Finally add the cream and mix well. Alternatively, put it all in the Thermomix and blitz on speed 5 for 20 seconds. Sift the flour with the baking powder and spices then mix this gently into the wet mix.
6. Pour the batter over the figs and place in the oven for 45 minutes. Check for to see if cake is cooked with a clean skewer. Insert it into the centre of the cake; if it comes out clean, remove from oven and allow to rest for 10 minutes. If not, cook for another 10 minutes and test again. Turn out onto a wire rack and quickly back over onto its base on another rack to cool.

A little cream or plain yoghurt is all that's needed to accompany this tender and moist little slice of the south.

Roti orloff

Pork is eaten much more frequently in Europe than it is here in Australia, particularly by the Germans. Steffi, one of our German woofers, introduced us to this rich pork dish, which looks complicated but is easy and quick to prepare. However, it actually originated in Belgium, so it's more French than German, one could say.

SERVES 6
Time 1½ hours

1kg or more of pork loin
Salt and pepper
10 slices of cheese, such as Gouda
10 slices of ham or smoky bacon
200mL cream

1. Set oven to 160°C.
2. Season the pork with salt and pepper and roast in a deep sided baking dish for 50 minutes.
3. Remove from oven and make 10 cuts into the meat about 2cm apart. Do not cut all the way through to the bottom of the meat – about ¾ of the way is fine. Note that the meat will not quite be cooked all the way through at this point.
4. Place a slice of cheese and piece of ham or bacon into each cut. You can truss with kitchen twine or just leave it as is.
5. Turn the oven up to 180°C. Pour the cream over the stuffed pork and return the dish to the oven to cook for a further 10 minutes. The cheese should be melted and starting to brown, and the pork should be cooked through.
6. Allow to rest 5 minutes, then cut the slices all the way through to separate them.

Serve with the cream sauce over mashed potato with a salad or green vegetables.

Steamed chicken Japanese style

This is one of those simple dishes that can be incredibly tasty if you use free-range or similar quality chicken. I'm not sure what the blanching does, but the result is just so Japanese. Hitomi, another gorgeous Japanese girl, cooked this for us a few years ago. She ended up following Jonathan, the French carpenter, around the world after meeting him while they were both woofing here at Oranje Tractor. They've been to some amazing places together since they were here. We really have travelled vicariously through our woofers.

SERVES 4
Time 25 minutes

4 chicken thighs, either whole, bone in, or cubed if the bone is removed
1 tablespoon grated fresh ginger (do not substitute powdered ginger)
1 or 2 cloves of garlic, crushed or finely chopped
4 tablespoons sesame oil
400mL rice wine
2 spring onions, finely sliced
1 tablespoon soy sauce
1 sprig of basil

1. Scald the chicken by placing in a pot of boiling, salted water for 3 minutes.
2. Remove, drain and pat dry with paper towel.
3. Heat the sesame oil in a frying pan and sauté the garlic and ginger over gentle heat. Do not allow the garlic to brown.
4. Add the chicken and sauté, then pour in the rice wine, spring onions and soy sauce. Cook for 5 to 10 minutes, or until the chicken is done.
5. Taste and add more soy sauce if desired, then add basil leaves and cook 1 more minute.

Great served with rice, of course, and some steamed greens.

Austrian apple strudel

This is one of Austria's most famous desserts, apart from Linzer Torte perhaps! Oliver, one of the few Austrians we've hosted, had a keen interest in all things food and wine, and made this version for us. I wouldn't have thought that filo pastry was particularly traditional, but it works very well. We visited Austria a couple of years after Oliver's stay with us, and were treated to his fabulous hospitality and tour guide-like services in the wine regions of this beautiful country. Of course, we enjoyed the locally produced version of apple strudel almost as much as Oliver's version. Although the culture and food are quite Germanic, the Austrians do have their own gastronomic culture. One custom that we particularly loved was that of the Heurigen. These wine-bar style venues in the wine regions in and around Vienna are similar to an Australian winery's Cellar Door. The laws permit only wine made by that producer in that year to be sold, and only for a limited period of time, usually a few months. Heurigen indicate that guests are welcome by hanging a couple of pine twigs above the entrance door. We were lucky enough to visit a few with Oliver to gain a good perspective on the concept. Needless to say, we loved it!

SERVES 6
Time 45 minutes

Filo pastry (it's the closest thing to the real struedel pastry you can get unless you make your own)
30g butter, melted
4 apples
120g breadcrumbs
50g butter, extra
50g sultanas (or currants or raisins)
½ teaspoon cinnamon
1 teaspoon grated lemon rind
Sugar

1. Set oven to 200°C. Line a baking tray with non-stick baking paper.
2. Heat some butter in a frying pan and gently fry the breadcrumbs until toasty. Allow to cool.
3. Peel and core the apple then chop into small cubes. Place in a bowl with the sultanas, cinnamon, about 1 tablespoon of sugar and the lemon rind.
4. Unroll the filo pastry and take four pieces out. Lay one piece on a well-floured board and brush it with cool melted butter. Put another piece on top, then sprinkle two tablespoons or more of the breadcrumbs down the centre of the pastry, lengthways.
5. Strain any liquid off the apple mix and place a layer of apple on top of the breadcrumbs. Brush one long edge with butter and begin rolling the other edge of the pastry over the apple to form a package. Tuck the ends under or seal them with more butter. Place carefully on the baking tray and repeat with the other two pieces of filo. Brush the tops with melted butter then bake for 30 minutes or until they are golden brown. Remove and when cool, dust with icing sugar.

Double cream goes perfectly with this.

Cumquat preserve

Keiko was a very sweet Japanese girl who shared this very sweet, traditional Japanese New Year speciality. Cumquats are a symbol of prosperity in Japan, so the plants are often given as gifts at this important time of the year. Their small, citrusy fruit are edible in their whole, raw form and have a sweet-sour characteristic due to the sweetness of the skin contrasting with the slightly sour flesh. In my opinion, the best marmalade is cumquat marmalade, but it is very time consuming to make, as cumquats have many, many seeds that need to be removed before cooking. Preserving them whole, as the Japanese do, therefore works perfectly for me. Keiko told us that Japanese women prepare these treats, along with a selection of others, so that when it comes to the Lunar New Year they too can enjoy the festivities without having to cook on the day. There are plenty of ways you can use these moreish morsels, from garnishing desserts to incorporating in salads – try Tuscan Kale with Cumquats – or simply enjoying them with a dollop of double cream or plain yoghurt.

SERVES MANY
Time 30 minutes plus 1 month

About 2kg Cumquats

Sugar Syrup
500mL water
250mL white wine (sweet or dry, but not oaked)
750kg white sugar
1 teaspoon grated fresh ginger

1. Place the cumquats in a pot and cover with water. Place on the heat and bring to boil.
2. Remove from heat and strain off the liquid. Refill pot with cold water and bring to boil again.
3. Make syrup by putting the water, wine, sugar and ginger into a large pot and heating, stirring until sugar has dissolved.
4. Tip the cooked cumquats into the syrup and bring to boil. Cook for 5 minutes then allow to cool slightly before pouring the cumquats and sugar into a large sterilised glass (heatproof) jar or ceramic pot. Ensure the jar is filled to the brim with syrup.
5. To sterilise the jar, put a large metal spoon into the jar and pour boiling water over spoon and into the jar.
6. Carefully rotate the jar to ensure water coats all sides. Empty the jar and allow to drain before using. Alternatively, heat in the oven at 100°C for 10 minutes.
7. Place a lid on the jar and store it for at least one month before eating.

Apple tarte tartin

Michelin Max was one of our more recent woofers. A young French chef with an impressive resume, Max demonstrated his cooking skills with much finesse but not a lot of fuss. His Michelin-star experience was clearly evident in everything he cooked. Well, at least that's what it looked like in the photographs Murray was sending me while I was on my 'Champagne Sabbatical' recently. At the time, I was being wined and dined in Champagne for the 40th Anniversary of the Vin de Champagne Award, Murray was at home, minding the fort and having his own gastronomic experience thanks to Max and his mate Antoine – also a former Michelin-starred chef. Although there are many exquisite dishes that could have been included here, most of them were complex or required special equipment or ingredients. Tarte Tartin is as French as fromage, is easy to make, and is sublime to eat. After crepes, this dish is the most often cooked recipe by our French guests. Sometimes a sweet pastry (pâte sablée) is used, and sometimes pâte brisée (the non-sweet pastry) is used. They have all been delicious, but I do prefer the brisée version so that there is a good foil for the very sweet caramel.

SERVES 6
Time 1½ hours

Pastry
200g flour
4g salt
100g unsalted butter
50mL water or milk

Apple layer
100g unsalted butter, chopped roughly into cubes
250g sugar
1.2kg apples (cooking apples are not necessary or even desirable)

1. To make pastry, mix salt (use a very small pinch if you don't have accurate scales) and the flour together then add the butter. Using your fingertips, rub butter into flour until it resembles fine breadcrumbs. Add the water or milk and combine until a firm dough is formed. Shape it into a smooth ball, cover with cling wrap and refrigerate for 30 minutes.

2. Make the caramel in a frying pan that can be put in the oven, or make it in a pot but be prepared to transfer the caramel to an oven proof dish. Put the sugar and a tablespoon of water into the pan, stir to combine then heat gently until the sugar melts and the colour changes. Take off heat and add a teaspoonful of butter, return to heat and stir. Continue adding the butter in teaspoonful until all of it is included. Turn off heat.

3. Set oven to 180°C. Peel and core apples then cut into halves. Turn the heat back on the pan of caramel and place the apples, cut side up, covering as much of the pan as you can. Allow to cook a few minutes.

4. If you used a pot, put the apples in and roll them around the caramel, then put the whole lot into a baking dish, turning the apples so that the cut side is up.

5. Now roll out the pastry to fit snugly over the apples in the pan or dish. Press down on the pastry and try to get the borders under the apples.

6. Bake for 30 minutes. Remove and allow to cool before inverting the tarte onto a plate.

Voilà! Now, enjoy your tarte tartin with cream or ice cream, as you wish.

Anglesey eggs

Chris, who preferred to be called David, was a proud young Welshman who left us with a Welsh cookbook and a host of other Welsh goodies as a parting gift. Although he enjoyed food, he was one of the few woofers who didn't cook. We didn't mind, however, because he always did such a good job of the washing up! He did supervise the creation of one of his mother's dishes, Anglesey Eggs. With a dish such as this it is important to use free-range eggs as they generally have so much more flavor, but for ease of peeling don't use the freshest ones you have. We are always looking for ways to use up the eggs our chooks give us, and this is a pleasant change from frittata and omelet as a quick and easy light meal. Freshly dug spuds and leeks enhance the final product even further. The original recipe requires Caerphilly cheese – a firm, white, salty cheesethat is a Welsh specialty – but it works well with the mixture of feta and mozzarella. We sometimes add some chopped anchovies for an extra blast of salt, but they are not essential, nor are they authentic to this Welsh recipe.

SERVES 4
Time 45 minutes
500g potatoes, peeled and chopped into large pieces
3 leeks, well washed and sliced thinly crosswise
6 eggs
600mL milk
50g butter
50g flour
50g feta cheese, crumbled
50g mozzarella cheese, grated
Salt and pepper to taste

1. Set oven to 200°C.
2. Boil a pot of water and add a teaspoon of salt. Put the chopped potatoes in and boil gently for 15 minutes or more, until they are soft. Drain the water from the potatoes.
3. In the meantime, braise the sliced leek in a pot with a little water until soft. Put the eggs into a pot of cold water and bring to boil, and simmer for 10 minutes. Remove the eggs immediately and put them under cold running water to cool.
4. Mash the potatoes roughly,then add the leeks, salt and pepper to your liking. Shell the eggs and cut them into halves, lengthways.
5. Make a white sauce by putting the milk in a saucepan with the butter and flour and over gentle heat stir continuously with a wire whisk until it is smooth and thick. Remove from heat, add the feta and stir gently. Season with salt and pepper.
6. Make a bed with the mashed potato and leek mix in a large oven proof casserole dish and arrange the egg halves, cut side up, on top. Pour over the white sauce and finish with the mozzarella cheese.
7. Bake in the oven for 15 minutes, or until it is golden on top.

Swiss choc mousse

This recipe breaks all my preconceived notions about not letting chocolate mix with water, and yet, by some miracle of food science, it works! Anna, one of the few Swiss visitors (sadly, there is no reciprocal working holiday travel visa arrangement between the Swiss and Australian governments), made a lasting impression with this and her suggestion for coating the 'lantern berries' (Anna's name for cape gooseberries) in chocolate. Not only did they look spectacular, of course they tasted divine. The following recipe is Anna's unconventional recipe for chocolate mousse.

SERVES 4
Time 25 minutes

100g good quality dark chocolate (70% cocoa)
Vanilla or honey or rum or grated ginger to flavour
2 teaspoons gelatine powder
250mL cream
1 tablespoon icing sugar

1. Place chocolate in a bowl then pour very hot water over it to cover. Allow to melt (stick a knife into the chocolate to test) until soft – do not stir!
2. Pour off the water then stir the chocolate until it is creamy.
3. Add a small amount of your flavouring of choice and stir again.
4. Prepare the gelatine by placing it in a small heat-proof cup of cool water. Stir and let it 'bloom' for 5 minutes.
5. Gently heat the cup in a small pot of water until the gelatine has dissolved. Put aside to cool down slightly while you whip the cream until it is thick.
6. Add the sugar to the cream then mix in the chocolate and gelatine.
7. Pour into serving glasses and chill for a few hours before serving.

SERVES 4 TO 6
Time 1 hour

Dough
450g flour
3 eggs
Pinch salt
1 tablespoon cool water

Filling

1 cup breadcrumbs (fresh, not toasted)
1 leek, sliced thinly
150g spinach, finely sliced and wilted (pour boiling water over) or frozen and defrosted
2 rashers of smoky bacon (or equivalent of speck), finely chopped
½ white onion, finely chopped
600g minced meat (pork, beef or combination of the two)
2 eggs
Pinch of salt
¼ teaspoon grated nutmeg
½ teaspoon marjoram
Pepper according to taste
1 whole onion
1 tablespoon olive oil
3 stems parsley
1 large pot of meat stock or salted water

German ravioli

One of our all-time favourite female German woofers, Sabine, introduced us to the joys of Maultaschen, which are rather like the Italian ravioli. Traditionally these noodle-dough pockets contain bacon, mustard and minced meat. When they are eaten in Swabia (the region containing Stuttgart, where Sabine lives) on Holy Thursday, the day before Good Friday at the Christian Easter, they have spinach added to them. According to Sabine, the Swabians enjoy these so much that every Thursday is Maultaschen day! We visited Sabine in her home town a year or so after her visit to Australia. She was very generous with her time, and as well as hosting us she also drove us through the Swabian countryside to our next destination on the German-Swiss border. Along the way we visited a perfectly renovated ancient farm, complete with real, old-fashioned beehive-shaped beehives. You know, the ones that the hairdos were named after! Germany has much to offer the visitor, and is sadly overlooked as a destination for travellers – apart from the Oktoberfest. Spectacular scenery abounds, great wine awaits and an interesting cuisine is available to enjoy. Germany is not just good for its beer festivals, either. They do wine festivals very well too. A few years ago, we were lucky enough to time our visit to a small town along the Mosel River such that it coincided with the Bernkastel Harvest Festival. This annual, well organised, well patronised event seemed to be the epitome of German food and wine culture. Every second stall along the cobbled streets of this picturesque, ancient town offered food and almost all of it was local specialities. No generic, bland, everyday tucker here! Potato cakes with apple sauce; Bratwurst in pretzel rolls; apple strudel (of course!) and Zwiebelkuchen. There were fireworks; street parades of the local wine producers; crowning of the Queen of the Vintage and much, much more. Closing time was in the early hours of the morning, when the last of the festival patrons left the streets. And this was just day one! Ah, they not only know how to work hard, the Germans party hard too.

1. Make the dough by combining the flour and salt in a large bowl. Mix the eggs and water together in a small bowl, then add this to the large bowl. Combine and knead until a firm dough is formed. Wrap in cling film and allow to rest in the fridge for at least half an hour.

2. Fry the onion, leek and bacon in the butter until softened, but not browned. Allow to cool. Put the breadcrumbs into a bowl and pour cold water over. Drain after a few minutes and squeeze excess water out. Add all filling ingredients to a large bowl and mix well until combined.

3. Remove dough from fridge, unwrap and cut into 4 pieces. Take one piece at a time to roll out. Keep the remainder wrapped. Use a pasta roller or a rolling pin to create a rectangular shape about 4mm thick. Keep your roller and your board well-floured to prevent sticking. Put about 1 tablespoon of filling mix on the dough every 8cm or so, down the middle of one side. Fold the other side over the filling, pressing down to form pockets. Ensure you press around each spoonful of filling to firmly encase it before cutting, with a pastry wheel or similar, into squares. Repeat rolling, filling and cutting process until finished.

4. Bring a large pot of salted water or, preferably, some meat stock to a boil. Add as many pockets to the pot as will comfortably fit and allow to simmer (not boil) for 15 minutes. Remove and keep covered while you repeat the cooking process.

5. In the meantime, slice another onion and chop some parsley. Fry the onion in olive oil over moderate heat until it starts to brown.

To serve, put a few pockets into each serving dish (soup bowl is ideal) and ladle some of the water or stock in to almost cover the pockets. Finish with fried onion slices and chopped parsley. These are often eaten with a German potato salad.

Jewish apple cake

This recipe has become a cellar-door favourite, and a woofer favourite too. It comes from the USA via Juliana, who contacted her Mom whilst she was here to request the recipe to help us deal with our glut of apples. It is remarkably easy and incredibly delicious. It took quite a while for it to dawn on me why this cake is called Jewish Apple Cake. Perhaps if it had been called Kosher Apple Cake I might have twigged sooner! There's no butter in the cake; instead, oil is used to give this cake its moist, delicious taste and texture. It is such an easy cake, dare I say foolproof? You can cut down the sugar a bit if you like as you get some sweetness from the apples, but I wouldn't drop it to less than $1\frac{1}{2}$ cups. Don't worry, it's a big cake and unless you eat it all yourself you will not be eating that much sugar. But if you're really concerned, you can replace the sugar with 1 cup dates (deseeded) and 1/2 cup raw honey. If you plan to use self-raising flour instead of plain, omit all but 1 teaspoon of baking powder. The cake will last for days, unless it gets gobbled up beforehand. As this is an American recipe, the ingredients are mostly listed in cups. It matters little whether you use American standard cup measures or metric cup measures, as it is the ratio of sugar, oil and flour that is important. I've listed the approximate weights in grams, but for this recipe it is probably better to use cup measures.

SERVES 10
Time $1\frac{1}{4}$ hours

4 eggs
2 cups sugar (or 380g)
1 cup oil (or 160g)
2 cups plain flour (or 320g)
½ teaspoon salt
1 teaspoon cinnamon
4 teaspoons baking powder
2 teaspoons vanilla
3 apples, peeled and sliced
1 cup walnuts, chopped (or 100g)
2 tablespoon sugar
1 teaspoon cinnamon

1. Set oven to 160°C.
2. In a large bowl, beat eggs, then gradually add the sugar and the oil.
3. In a separate bowl, sift the flour, salt, cinnamon and baking powder. Fold the flour mixture into the egg-sugar mixture. Add vanilla, apples and walnuts and stir briefly to combine.
4. Grease and flour a cake tin then add the cake mixture. Mix together the remaining sugar and cinnamon, and sprinkle this on top of the mixture. Bake 50 or 60 minutes.

Zopf

Zopf is a traditional bread from Switzerland and has been since the middle of the 15 century, according to Joelle one of our Swiss woofers. She said that its origin comes from a custom whereby widows cut off their braided hair and buried them with their husbands. Later on they buried a bread loaf in this shape instead of their hair.

Switzerland may not be seen by many as a unique gastronomic paradise, yet the Swiss do have their traditional foods and many of them, like this one, are particularly tasty. One of my first and fondest memories of food in Switzerland was when I was a young backpacker travelling in Europe. I went to visit my mother's best friend, an Aussie who'd married a Swiss man. She kindly offered me lunch, something special that she'd cooked up for me – asparagus. I was a little horrified because the only asparagus I had ever eaten at that tender age was tinned asparagus, and was repulsed by the soft, squishy texture and overcooked flavour. But being bought up to never turn up my nose at food being cooked for me by others, I took a deep breath and waited for the lunch to be served. Imagine my surprise when a plate full of fresh, steaming asparagus spears bathed in Swiss butter appeared before me. My horror abated, especially once I tasted my first, fresh asparagus spear. I was now an asparagus fan for life.

SERVES 6 TO 8
Time 2 hours

500g flour
2 teaspoons salt
1½ teaspoons yeast
1 teaspoon sugar
250mL milk
50g butter
1 egg

1. Prepare a tray by lining it with baking paper.
2. Mix the flour, salt, yeast and sugar together in a bowl.
3. Melt the butter and mix it with milk, then add this to the dry mix and form into a dough.
4. Place the dough into a clean bowl and put a clean, damp tea towel on top and let it rise for about one hour.
5. Remove the dough and cut into 3 equal pieces, forming 'snakes', approximately 10cm long.
6. Pinch the 3 snakes together at one end and then plait the snakes to form the Zopf, pinching again to close the plait.
7. Make an eggwash by cracking the egg into a bowl and whisking lightly.
8. Brush the Zopf with the eggwash twice, then place it on the tray and allow to rest 5 minutes.
9. Place into the (cold) oven and turn it on to 200°C.
10. Cook for 40-50 minutes.

Orange & almond cake

This is one of my favourite cakes to eat and to make. It is also one that gets rave reviews from our guests – both the paying type and the woofing type. I usually boil up one big pot of oranges at a time, then purée and freeze them for ease of use later in the year. Having frozen orange purée on hand means that this is a quick and easy, and mostly fool-proof, cake to whip up – just don't get the amount of baking powder wrong. Trust me, you will have an orange Mt Vesuvius on your hands, or at least on your oven floor. If you use pre-ground almond meal you will get quite a different texture from what you see here. I use the Thermomix to grind the almonds for about 20 seconds to get a larger textured meal, and most other blenders could probably give the same result. You can use stevia instead of sugar if you need this gluten-free cake to be sugar-free too. You can also spice it up – cardamom is a lovely, Middle Eastern influence – but it is great just as it is. If you don't have fresh or frozen orange purée on hand, you can take a short cut: use a 300g jar of orange marmalade instead of the oranges and sugar. It is a much stronger flavoured cake but still delicious.

SERVES 8
Time 45 minutes (plus cooking time if using fresh oranges)

2 whole oranges (or about 300g frozen orange purée)
Water to cover
300g ground almonds
200g sugar
4 eggs
10g baking powder (gluten free)

1. Prepare a round cake tin with non-stick baking paper or by greasing well.
2. Put the oranges in a large pot and cover with water. Bring to a boil then simmer for 1 hour.
3. Remove the oranges and allow to cool. Cut them in half cross-wise and remove any seeds.
4. Purée in a blender until uniform but not smooth.
1. Set the oven to 180°C.
2. In a large bowl, mix the eggs and sugar until combined. Add the ground almonds, puréed oranges and baking powder and mix again.
3. Pour the cake batter into the prepared pan and place in oven. Cook for 45 minutes and check for doneness by inserting a skewer. If cooked, the skewer will come out clean. If not, cook for another 10 minutes and check again. If so, remove from oven and allow to cool in the tin. Turn out onto a rack, then onto a serving plate.

A simple dusting of icing sugar is all that is needed to garnish. Serve with plain yoghurt or whipped cream.

Sardines in vine leaves

Sardines, being very much a local product in Albany, have featured frequently on the Oranje Tractor menu. Early in the peace, Flavien – a French chef wwoofing with us – suggested a minor modification to the offering and transformed the fish from good to great. Rather than simply pan-fry and adorn with salt, black pepper and lemon juice, Flavien's method involves using grape vine leaves, à la the Greek Dolmades, and incorporating thyme leaves and white pepper. This preserves the moisture in these little fish and adds great flavour and texture. You can buy preserved vine leaves, but I don't bother. If you have access to a grapevine (and the permission of its owner), preparing your own fresh leaves is easy and results in a superior end result. You need to pick young leaves as older ones will be fibrous and tough. The youngest leaves are those at the very tip of the growing canes. Take the fourth and fifth one along each vine, so you get decent sized leaves that are still tender. It will not harm the vine or the grapes. Depending on the variety, and hence the size of the leaves, you might need to take the third and fourth ones instead to get leaves that are not so big that they wrap around the fish several times. Pick enough for the number of sardines you have, plus 10% to allow for tearing. Put the leaves, without any stem, into a metal bowl and pour boiling water over. Allow them to sit for 5 minutes or until they have all turned from bright green to olive green. Drain and pat dry.

SERVES VARIES
Time 20 minutes

Sardine fillets
Grapevine leaves
Thyme leaves
White pepper
Olive oil
Lemons

1. Take your sardines and split them into two fillets, if not already prepared like this. You may want to cut off the dorsal fin, but it is not essential.
2. Lay the fillet lengthwise down the centre of a leaf, which should have its back side (the veiny side) facing you. Put a pinch of thyme leaves on the fish and sprinkle with a little white pepper.
3. Fold the leaf over, and tuck under, so that you have a little green parcel of fish.
4. Once all fillets are assembled, heat a frying pan with olive oil. Fry each vine-wrapped fillet for only 30 seconds on each side.
5. Remove, drain on kitchen paper, dust with a few salt flakes and squeeze a bit of lemon juice over.

Served with garlic mayonnaise, these humble fish are elevated to a new height of culinary enjoyment. Alternatively, these are great on the barbecue in summer.

Zito

Our enthusiastic and entertaining Slovenian repeat woofer, Marko, introduced us to this rustic combination of wheat and honey. Marko had an excellent command of the English language and had plenty to say. A man of great intellect and creativity, Marko's past work as an architect and as a traveller extraordinaire gave him plenty of grist for the mill to share with us. We enjoyed many great discussions with him... when he wasn't working in the vineyard! And he worked even more than he talked, with even greater gusto! He single-handedly, whipper-snipped underneath the 60 or so vine rows in just one week. Phew, we were exhausted watching him work as the sun slipped away each evening.

SERVES 2
Time 1 to 12 hours

1 cup whole wheat grains
1 tablespoon honey
1 tablespoon walnut pieces

1. Put the wheat into a pot and either pour boiling water over to cover and leave overnight (12 hours) or put cold water over and bring to the boil.
2. Simmer for an hour, until the grains soften, then strain.
3. Heat the wheat grains in a small pot with the honey and add the walnut pieces.

Serve with cream or plain yoghurt.

Spring 2010 Literary Lunch with Emmanuel Mollois

Aperit - served with OT Sparkling Riesling 2006

- *s* of Bouverie smoked trout bruschetta on rounds of toasted brioche (Bistro Dupont recipe)
- herbed palmiers (adapted from et Voila)

f

Entre – served with OT 2007 Rose

Charcuterie plate of chicken liver pate, pate encroute (pastry from et Voila), and rabbit rillets
d platter for 4 with pickled oranje tractor grapes, gardener's kitchen pickled cherries,
& cornichons, accompanied by Royale breads

f

OT 2005 Riesling or 2008 Shiraz

white wine jus

Epilogue

Writing this little memoir required me to pause and think. This episode of reflection, consequently, has enabled me to better understand why this tractor life we live is not a typical life. For the past eight years I have not had to endure the rigmarole of dealing with a bureaucracy, of having to leave our lovely home to spend eight hours at work in a bland office, nor of being asked to increase my output without any more inputs. Murray says he no longer has to deal with 'administrivia', which sums it up quite well, I think. Hosting travellers under the WWOOF scheme has also reinforced the notion of 'living in the moment'. Our guests really do seem to 'be present' and enjoy even the most basic tasks and what we sometimes take for granted - our unique environment. We are truly blessed to live in such a beautiful part of Western Australia, and I realise just how lucky we are.

This is not to say that this tractor life is carefree and easy – far from it. But, when one does stop and take stock, the benefits far outweigh the costs. Sure, we don't have regular weekends most of the year; we can't buy the latest clothing fashions each season or new cars every few years; and we can't lounge about during our time off because we don't really get any. But we are, largely, the masters of our own destiny; we only have to answer to each other; we spend most of our time doing practical, creative things that bring joy to others as well as ourselves; and we do it on our lovely patch of this earth - surrounded by wonderful friends and family, a supportive community and Merlot the dog!

The Last word

On many occasions I have had the pleasure of eating food. All manner of food from all manner of cultures. Sometimes on a plate and sometimes off a napkin. I do love a mix of food and cultures. And I do love a regular visit to Oranje Tractor, my favourite winery, indeed, the only winery I regularly visit, mainly because I don't drink wine. But this place is much more than wine, as is evidenced by this fabulous book.

It is also a place to hang out, chat with friends, enjoy fine music, watch a movie, or take up an implement. I have taken a pair of grape harvesting snips and set to it with gloves and enthusiasm, cutting away the dead and unwanted. By morning tea I was reliably informed that I had broken the record for the number of finger cuts for the period. Then by lunch I was informed, and again by afternoon tea and, finally, at the end of the day, when all fingers were counted, it seemed I had cut more fingers than I had to hand. My name is deeply etched on the finger cutting honour board and if you buy a wine of that year you can safely assume my blood is in your glass.

But back to the reason I am here, this book. The author, Pam Lincoln, is known to me, well enough for me to say things about her that I will not repeat here, other than to say that many of them include the word love. If she asks me to eat with her, at her place, her cooking, I attend, whatever my personal circumstances. Indeed, my partner and I contributed to the crowd funding campaign for this handsome volume. Would you believe I have tried all the recipes and each and every one of them is a masterpiece of culinary creativity? How about I have attempted to cook each and every recipe in this book and failed each and every time? Why so many failures? Something in the ingredients? The instructions? No, nothing could be further from the truth. It is simply because I, unlike Pam Lincoln, am an incompetent goose in the kitchen and I only ever cook when my partner leaves the house and freezer empty. I think she is trying to tell me something. Actually, none of that is true. There is one certain truth, however, and that is that I delight in exotic foods, especially those not from my town and very few of these are. *This Tractor Life* is full of recipes from people who have travelled from all over the world to share a plate on a plot of paradise, in an isolated and glorious region of planet earth.

If you visit my house you will find *This Tractor Life* front and centre on our cook-book shelf and all who enter the kitchen are instructed to use it, or eat takeaway.

Pam, I love you.

Jon Doust

author
eater

Alphabetical

- Anglesey eggs 171
- Apple butter 125
- Apple tarte tartin 169
- Asparagus panna cotta 75
- Austrian apple struedel 165
- Avocado-coffee smoothie 110
- Black bean houmous 59
- Blaukraut 87
- Caramelised sweet potato soup 139
- Choko wedges 65
- Cinnamon rolls 121
- Crème citron 93
- Cumquat preserve 167
- Cyrielle's killer chicken 71
- Deep fried brussels sprouts 74
- Eldas fresh green olives 135
- Finnish farmer biscuits 105
- German ravioli 175
- Greek-style stuffed summer vegies 90
- Hannah's apple bake 145
- Japanese chicken curry 137
- Jeff's chilli sauce 89
- Jewish apple cake 177
- Lemon and rosemary cake 91
- Max's grandma's french fries 109
- Mirror cake 99
- Mona's backpacker muffins 69
- Okonomiyaki: japanese vegetable pancakes 61
- Olive cake 129
- Orange and almond cake 181
- Paleo chocolate cake 67
- Pancake cake 109
- Pastiera napoleatana: italian easter cake 55
- Pear and passionfruit flan 63
- Pear and almond soufflé 143
- Pear, raspberry and chocolate crumble 127
- Pepernoten: spiced dutch biscuits 103
- Pflammenkuchen: german plum cake 101
- Potato dauphinoise 53
- Prawns with garlic and grass fines 51
- Provencal fig cake 159
- Quince baked, middle eastern style 128
- Rote gruetze: german red fruit dessert 131
- Roti orloff: roast pork with cheese 161
- Sardines in vine leaves 183
- Scottish mess 97
- Smoked trout whip 57
- Snails, southern france style 73
- Spaetzle: german noodles 141
- Spicy pear chutney 133
- Steamed chicken, japanese style 163
- Susan's famous "best ever" cheesecake 95
- Swiss choc mousse 173
- Zito: a hearty slovenian brekky 184
- Zopf: swiss breakfast bread 179
- Zwiebelkuchen: german harvest cake 123

Type

FISH/SEAFOOD
- Prawns with garlic and grass fines 51
- Sardines in vine leaves 183
- Smoked trout whip 57

FRUIT
- Apple tarte tartin 169
- Austrian apple struedel 165
- Greek-style stuffed summer vegies 90
- Jewish apple cake 177
- Mirror cake 99
- Orange and almond cake 181
- Pear and passionfruit flan 63
- Pear, raspberry and chocolate crumble 127
- Pflammenkuchen: german plum cake 101
- Provencal fig cake 159
- Rote gruetze: german red fruit dessert 131
- Scottish mess 97

GLUTEN FREE
- Asparagus panna cotta 75
- Avocado-coffee smoothie 110
- Black bean houmous 59
- Caramelised sweet potato soup 139
- Crème citron 93
- Cumquat preserve 167
- Cyrielle's killer chicken 71
- Eldas fresh green olives 135
- Hannah's apple bake 145
- Japanese chicken curry 137
- Lemon and rosemary cake 91
- Mona's backpacker muffins 69
- Paleo chocolate cake 67
- Pear and almond soufflé 143
- Pflammenkuchen: german plum cake 101
- Potato dauphinoise 53
- Quince baked, middle eastern style 128
- Roti orloff: roast pork with cheese 161
- Scottish mess 97
- Snails, southern france style 73
- Spicy pear chutney 133

MEAT
- Cyrielle's killer chicken 71
- German ravioli 175
- Steamed chicken, japanese style 163

SAUCE
- Apple butter 125
- Jeff's chilli sauce 89

VEGETARIAN
- Anglesey eggs 171
- Black bean houmous 59
- Blaukraut 87
- Choko wedges 65
- Deep fried brussels sprouts 74
- Okonomiyaki: japanese vegetable pancakes 61
- Potato dauphinoise 53

Main ingredient

APPLES	Austrian apple struedel	165
	Hannah's apple bake	145
	Jewish apple cake	177
AVOCADO	Avocado-coffee smoothie	110
BERRIES	Mirror cake	99
	Scottish mess	97
BLACK BEANS	Black bean houmous	59
BRUSSELS	Deep fried brussels sprouts	74
CABBAGE	Blaukraut	87
CAPSICUM	Greek-style stuffed summer vegies	90
CHEESE	Susan's famous "best ever" cheesecake	95
CHICKEN	Cyrielle's killer chicken	71
	Japanese chicken curry	137
	Steamed chicken, japanese style	163
CHILLIES	Jeff's chilli sauce	89
CHOCOLATE	Mona's backpacker muffins	69
	Paleo chocolate cake	67
	Swiss choc mousse	173
CHOKOES	Choko wedge	65
CUMQUATS	Cumquat preserve	167
EGGS	Spaetzle: german noodles	141
	Anglesey eggs	171
	German ravioli	175
	Okonomiyaki: japanese vegetable pancakes	61
	Pancake cake	109
FIGS	Provencal fig cake	159
FLOUR	Finnish farmer biscuits	105
	Pepernoten: spiced dutch biscuits	103
	Zopf: swiss breakfast bread	179
GRAPES	Rote gruetze: german red fruit dessert	131
LEMON	Crème citron	93
	Lemon and rosemary cake	91
OLIVES	Eldas fresh green olives	135
	Olive cake	129
ONIONS	Zwiebelkuchen: german harvest cake	123
ORANGES	Orange and almond cake	181
PEARS	Pear and passionfruit flan	63
	Pear and almond soufflé	143
	Pear, raspberry and chocolate crumble	127
	Spicy pear chutney	133
PLUMS	Pflammenkuchen: german plum cake	101
PORK	Roti orloff: roast pork with cheese	161
POTATO	Max's grandma's french fries	109
	Potato dauphinoise	53
PRAWNS	Prawns with garlic and grass fines	51
QUINCES	Quince baked, middle eastern style	128
RICE	Pastiera napoleatana: italian easter cake	55
SARDINES	Sardines in vine leaves	183
SNAILS	Snails, southern france style	73
SPICES	Cinnamon rolls	121
SWEET POTATO	Caramelised sweet potato soup	139
TROUT	Smoked trout whip	57
VEGETABLES	Asparagus panna cotta	75
WHEAT	Zito: a hearty slovenian brekky	184

Cuisine

AMERICAN	Jewish apple cake	177
	Susan's famous "best ever" cheesecake	95
AUSTRIAN	Austrian apple struedel	165
BELGIAN	Roti orloff: roast pork with cheese	161
CANADIAN	Black bean houmous	59
DUTCH	Pepernoten: spiced dutch biscuits	103
ENGLISH	Spicy pear chutney	133
FINNISH	Finnish farmer biscuits	105
FRENCH	Apple tarte tartin	169
	Asparagus panna cotta	75
	Crème citron	93
	Cyrielle's killer chicken	71
	Max's grandma's french fries	109
	Mirror cake	99
	Olive cake	129
	Pancake cake	109
	Pear and almond soufflé	143
	Pear, raspberry and chocolate crumble	127
	Potato dauphinois	53
	Prawns with garlic and grass fines	51
	Provencal fig cake	159
	Snails, southern france style	73
GERMAN	Apple butter	125
	Blaukraut	87
	German ravioli	175
	Hannah's apple bake	145
	Paleo chocolate cake	67
	Pflammenkuchen: german plum cake	101
	Rote gruetze: german red fruit dessert	131
	Spaetzle: german noodles	141
	Zwiebelkuchen: german harvest cake	123
GREEK	Greek-style stuffed summer vegies	90
INTERNATIONAL	Pear and passionfruit flan	63
	Avocado-coffee smoothie	110
	Caramelised sweet potato soup	139
	Choko wedges	65
	Deep fried brussels sprouts	74
	Jeff's chilli sauce	89
	Lemon and rosemary cake	91
	Mona's backpacker muffins	69
	Orange and almond cake	181
	Sardines in vine leaves	183
	Smoked trout whip	57
ISRAELI	Quince baked, middle eastern style	128
ITALIAN	Eldas fresh green olives	135
	Pastiera napoleatana: italian easter cake	55
JAPANESE	Cumquat preserve	167
	Japanese chicken curry	137
	Okonomiyaki: japanese vegetable pancakes	61
	Steamed chicken, japanese style	163
SCOTTISH	Scottish mess	97
SLOVENIAN	Zito: a hearty slovenian brekky	184
SWEDISH	Cinnamon rolls	121
SWISS	Swiss choc mousse	173
	Zopf: swiss breakfast bread	179
WELSH	Anglesey eggs	171

Meal

APPETISER
Asparagus panna cotta	75
Black bean houmous	59
Caramelised sweet potato soup	139
Hannah's apple bake	145
Olive cake	129
Sardines in vine leaves	183
Smoked trout whip	57
Snails, southern france style	73

BREAKFAST
Avocado-coffee smoothie	110
Cinnamon rolls	121
Zito: a hearty slovenian brekky	184
Zopf: swiss breakfast bread	179

MAIN
Anglesey eggs	171
Cyrielle's killer chicken	71
German ravioli	175
Greek-style stuffed summer vegies	90
Japanese chicken curry	137
Okonomiyaki: Japanese vegetable pancakes	61
Prawns with Garlic and Grass Fines	51
Roti orloff: roast pork with cheese	161
Spaetzle: german noodles	141
Steamed chicken, japanese style	163
Zwiebelkuchen: german harvest cake	123

SIDE
Blaukraut	87
Choko wedges	65
Deep fried brussels sprouts	74
Max's grandma's french fries	109
Potato Dauphinoise	53

CAKE/DESSERT
Apple tarte tartin	169
Austrian apple struedel	165
Cinnamon rolls	121
Crème citron	93
Finnish farmer biscuits	105
Jewish apple cake	177
Lemon and rosemary cake	91
Mirror cake	99
Mona's backpacker muffins	69
Orange and almond cake	181
Paleo chocolate cake	67
Pancake cake	109
Pastiera napoleatana: italian easter cake	55
Pear and passionfruit flan	63
Pear and almond soufflé	143
Pear, raspberry and chocolate crumble	127
Pepernoten: spiced dutch biscuits	103
Pflammenkuchen: german plum cake	101
Provencal fig cake	159
Quince baked, middle eastern style	128
Rote gruetze: german red fruit dessert	131
Scottish mess	97
Susan's famous "best ever" cheesecake	95
Swiss choc mousse	173

CONDIMENT
Apple butter	125
Cumquat preserve	167
Eldas fresh green olives	135
Jeff's chilli sauce	89
Spicy pear chutney	133

DRINK
Avocado-coffee smoothie	110

www.ingramcontent.com/pod-product-compliance
Lightning Source LLC
Chambersburg PA
CBHW061156010526

44118CB00027B/2995